KT-196-026

£17·50

WEATHER AND BIRD BEHAVIOUR

Weather and bird behaviour

By NORMAN ELKINS

Illustrated by
CRISPIN FISHER

SECOND EDITION

T & A D POYSER
Calton

ISBN 0 85661 051 8 60556552

First published in 1983
Second edition published 1988
by T & A D Poyser Ltd
Town Head House, Calton, Waterhouses, Staffordshire, England

Text set in 10/11½ pt Linotron 202 Plantin, printed and bound in Great Britain at The Bath Press, Avon

Contents

List of plates

List of figures

List of tables

Preface and acknowledgements

The stimulus for this book initially came from ornithological friends who were interested in the weather and curious about its effects upon the subjects of their studies. The idea developed hesitantly, but then more forcibly when it became clear to me that no-one had previously tackled the task. The result is, I hope, a readable reference work which will fill what appears to be a substantial gap in the literature.

I must first of all acknowledge a great debt to Professor George Dunnet and the Science Library of Aberdeen University. Without free access to consult references, the book could not have been written. A similar large debt is due to the staff of the National Meteorological Library at Bracknell, for assistance and loans of material. Many other individuals and organisations loaned or gave particular reference material, and of these I must especially thank Dr W. R. P. Bourne, the libraries of the British Trust for Ornithology and the Scottish Ornithologists' Club (whose librarian, Bill Harper, also provided certain photographs), and the Elgin Public Library. Translations from German and Dutch were kindly undertaken by Mrs I. McPhail and Mrs L. Hare respectively. Brian Etheridge permitted the use of his Crag Martin weight data used in Fig. 17, and Sandy Comfort provided photocopying facilities. A further debt is owed to Grace and Reg Bisset for unstinting service with a typewriter, and much general and stimulating discussion.

Drs Bill Bourne, Peter Evans and Ian Newton kindly read and commented upon certain chapters, and gave me the benefit of their expert advice. Thanks are also due to Bob Hudson for invaluable advice and suggestions for improvements to the text. Any errors are entirely my own. I hope few have crept in, and for those I beg the reader's forgiveness towards one who, although a meteorologist by training, pursues ornithology only as a hobby. Finally, I would like to thank my family for their patience and tolerance while the book was being written.

All meteorological data used in the figures are Crown Copyright and published by kind permission of the Controller, Her Majesty's Stationery Office. They are extracted from Meteorological Office publications and reports, including the Daily and Monthly Weather Reports, and the Daily Aerological Record. Figures 15 and 37 were originally published in the *Meteorological Magazine*, and Fig. 38 in the *Marine Observer*. Figure 2 is reproduced by permission of the American Meteorological Society; Fig. 12, with permission, from *Finches* by Ian Newton in The New Naturalist Series, 1972, published by William Collins, London; Figs. 18 and 40, and data in Figs. 39 and 42, by permission of the British Trust for Ornithology from their journal, *Bird Study*; Fig. 20, with permission, from *Swifts in a Tower* by David Lack, published by Chapman and Hall; Fig. 31 and Table 3, with permission, from the monthly journal, *British Birds*; Table 5 by permission of the Scottish Ornithologists' Club from their journal *Scottish Birds*, and Fig. 32, with permission, from *Population Ecology of Raptors* by Ian Newton, published by T. & A. D. Poyser.

PREFACE TO SECOND EDITION

In common with all scientific disciplines, there has been a proliferation of meteorological and ornithological literature since the first edition was completed and, in addition to minor textual revisions, I have therefore incorporated data on recent events and research. I am grateful to my wife, Jean, and my colleague, John Cubin, for translations from Spanish and German respectively.

Introduction

The majority of living creatures inhabiting the biosphere of this planet are affected in some way by the atmosphere and its changes. Almost all forms of life that use flight are influenced by atmospheric processes. Many insects and most birds spend much of their lives within the lower layers of the atmosphere and, indeed, changes in this environment are an integral part of the ecological complex in which birds live.

We are all aware of the more obvious meteorological problems that birds must overcome in order to survive. We can see how well gulls fly in a gale-force wind, with an ease that belies the buffeting to which humans are subjected on the ground, and we know how small birds flock to our gardens in frost and snow in the search for food. Most of us also realise that large numbers of birds perish on migration – but there are innumerable other, more subtle, consequences of variations in weather.

It is the purpose of this book to review the meteorological aspects of the avian environment, but before one examines in detail how the weather influences bird behaviour, one must look at the basic principles of weather generation. First, it is necessary to define the requisite expressions, to which end I quote precise definitions adopted at a recent World Climate Conference:[102]

Weather is associated with the complete state of the atmosphere at a particular instant of time, and with the evolution of this state through the generation, growth and decay of individual atmospheric disturbances.

Climate is the synthesis of weather over a period (at least a decade, and normally 30 years) long enough to establish its statistical properties, and is largely independent of any instantaneous state.

There are two further terms which have an important bearing on the subject – climatic change and climatic variability:

Climatic change may occur when the statistical properties of a succession of years or longer periods differ consistently and significantly from the normal.

Climatic variability denotes variations within the climate, shown by marked differences of a particular month, season or year from the corresponding long-term average, and fluctuations in statistical properties over periods of weeks, months or years.

Climate and climatic change (which are responsible for most types of habitat, and through this the distribution of animal life that inhabits a particular environment) are beyond the scope of this book. Suffice to say that many aspects of climatic change are little understood and often difficult to correlate with variations in bird populations. This holds particularly over large areas where data scarcity and analytic difficulties, not to mention interpretation of apparent changes, make accurate conclusions almost impossible.

I have therefore limited the material in this book to the description of the short-term consequences of climatic variability and the more immediate effect of weather. These factors have a profound influence upon the behaviour of individuals and of populations. Populations contain individuals which are able, in varying degrees, to cope successfully with a range of conditions in their physical environment. These conditions include temperature, humidity, airflow and rainfall; and marked changes outside the normal range are capable of endangering the survival or fertility of individuals less able to meet them. Behaviour varies in response to stresses on an individual. Periods of climatic variability may impose major burdens upon resident populations which cannot adapt to temporary but irregular (and perhaps prolonged) periods of severe conditions. Climatic variability embraces all the anomalies that imprint themselves on one's memory – in Britain, the severe winters of 1946/47 and 1962/63, the drought of 1975–76, and numerous shorter but memorable spells of distinctive weather. Since most birds live on or close to the earth's surface, it is also necessary to explore the micro-meteorological and microclimatological aspects of this region to increase one's understanding of avian behaviour.

The ultimate motive for most of a bird's activities is the finding of sufficient food to ensure survival and successful reproduction. Dr David Lack[87] maintained that starvation outside the breeding season is much the most important density-dependent factor in wild birds. It helps to balance losses against recruitment and so maintain a population at about the same density. An independent factor on the loss side of the equation is the weather.[195]

Food availability is undoubtedly a significant element in population control but, nevertheless, each species is adapted to take maximum advantage of the food supply in its preferred habitat. Food supply in general is closely regulated by climate, which controls the habitat in respect of vegetation and conditions in the earth's water bodies. Through its effect upon food supplies, climate also determines the mobility of individual birds. In mid and high latitudes, many bird populations show differences in mobility which are dependent partly on the degree of influence that the oceans exert on the climate of their habitat.

Differences in feeding behaviour from season to season are part of a bird's adaptation to variations in habitat and food supply related to seasonal climates. A feature of great importance throughout the year is weather, since it affects feeding in two ways. Firstly, it directly modifies the acts of feeding, foraging or hunting. Secondly, it influences the availability of vegetable foods and the behaviour and survival of animal prey, vertebrate and invertebrate. Therefore the degree to which the feeding success of an individual is influenced by weather is determined partly by its food requirement and partly by its method of feeding. For example, aerial feeders are affected largely by temperature and air movement; the food of seed-eaters is controlled partly by climatic variability; oceanic birds are relatively immune to weather except in certain wind conditions; and predators (in the widest sense) are influenced by the behaviour of their prey. If the predators eat carrion they may even benefit in the short-term from prey mortality in adverse weather.

In setting out to write this book I had the immediate problem of trying to marry two very different scientific disciplines – one physical and one biological – whilst still ensuring that the reader could grasp the more complex points of each. The task of explaining the relevant aspects of meteorology in a clear and concise form is difficult. I would ask the reader not to be daunted by what may seem a plethora of unfamiliar terms; I am sure that a careful study of the text and figures in Chapter 1 will lead to greater understanding and interest in subsequent chapters.

Whilst searching the literature it became clear that, apart from migration, little study had been made of weather and bird behaviour, and much relevant material was scattered through works on other ornithological topics. Owing to the vastness of the subject I have limited references chiefly to birds of the western Palearctic in Eurasia and Africa and, to a lesser extent, those of North America, and material has been gleaned only from the more readily-available publications. Because much of the material occurred in works on other aspects of bird behaviour, it was often difficult to detect precisely what particular meteorological conditions were involved. Many generalised terms, such as 'bad' weather, were used, possibly indicating the involvement of one or more elements of weather, which of course affect different species in different ways.

Throughout the book I have tried to bring constancy to units of measurement, using SI units except where conventional meteorological usage differs, e.g. pressure is measured in millibars (mbar) and temperature in degrees

Celsius (°C). In the case of wind velocity, I prefer to use metres per second (m/s) rather than kilometres per hour (kph). Both are probably unfamiliar to many readers, but m/s has the advantage of being easily converted into knots (nautical miles per hour) by multiplying by two. Conversions will be found in Table 2.

CHAPTER ONE

The weather

An understanding of the basic mechanics of atmospheric motion is necessary if one is to appreciate their influence upon birds' activities. Meteorology is a highly complex science bound by precise mathematical and physical laws. The difficulty with which measurements of the state of the atmosphere are made raises problems which militate against complete understanding. However, the formation of the elements that make up the weather is well enough understood, and in this chapter I will examine briefly the causes and effects of the various atmospheric processes, by considering a broad-scale description of the atmospheric circulation. To facilitate the understanding of relationships described later in the book, I will also define some of the rather specialised terms, and look at the values of the various meteorological elements that are found in an avian environment. Should the reader find that he requires further knowledge, more detailed treatment can be found in standard meteorological textbooks[97,178].

ATMOSPHERIC CIRCULATION

The earth's atmosphere is shallow when compared to the diameter of the planet. It is a gaseous fluid bound to the earth by gravity; a tenuous envelope three-quarters of which lies between the earth's surface and an altitude of around 12 km. The mass of the atmosphere above any given point exerts a pressure at that point, the pressure decreasing with altitude as the atmosphere becomes less dense. The concept of atmospheric pressure is one to which we will return shortly, since it is fundamental to meteorology.

The planet rotates at a more or less constant rate, once in 24 hours, about a polar axis which is tilted in relation to the sun. At the same time the earth revolves round the sun once per year. Its shape is roughly spherical, although slightly flattened at the poles. These three factors – the daily planetary rotation, yearly revolution and oblate spheroid shape – result in unequal amounts of the sun's radiative energy falling on the planet, with variations in time and space. The radiation, principally in the form of heat, provides the energy which maintains life on the earth, and its variation over the planet is the dominant cause of atmospheric change.

The spherical shape of the rotating earth affects the movement of air across its surface. Since the vertical axis of an observer standing at one of the poles is in line with the earth's axis, he will rotate about that axis at the same rate of rotation as the earth. An observer on the equator, with his vertical axis at right angles to that of the earth, has no such rotation. Thus, the degree of rotation about the local vertical axis varies with latitude, and a particle of air moving across the earth's surface in any direction (other than along the equator) would appear to an observer in the northern hemisphere to curve to the right, and in the southern hemisphere to the left. This apparent deflecting force is known as the Coriolis force, and its value increases polewards from zero at the equator to a maximum at the poles.

About 45% of the radiation from the sun passes through the atmosphere to be absorbed by the earth's surface. The heat is then re-distributed from below by various methods. One result is a fall in temperature with height; a decrease which more or less ceases at an altitude of between 7 and 12 km, although this height is greater nearer the equator. This level is known as the tropopause, and above it, in the stratosphere, there is little decrease in temperature with height, since a certain amount of the sun's radiation is absorbed by a layer of ozone. The remainder of the incoming radiation is absorbed by water vapour or reflected from clouds or snow.

The atmospheric layer between the tropopause and the earth is known as the troposphere, and its depth varies with temperature. Because of the almost vertical position of the sun, tropical regions experience the maximum amount of radiation per unit of surface area. Physical laws decree that, at a constant pressure, a mass of air expands on receipt of incoming heat. This results in a greater volume of air above the equator than nearer the poles; consequently a horizontal difference of pressure exists at high levels. The heating from the underlying surface is transferred upwards, not uniformly but within convection cells. Bubbles of air warmer than their surroundings, and therefore less dense and more buoyant, rise to the tropopause. To achieve a balance between this excess of equatorial air and the air in higher latitudes, a pole-ward flow of air is set up in the upper troposphere. This flow is deflected

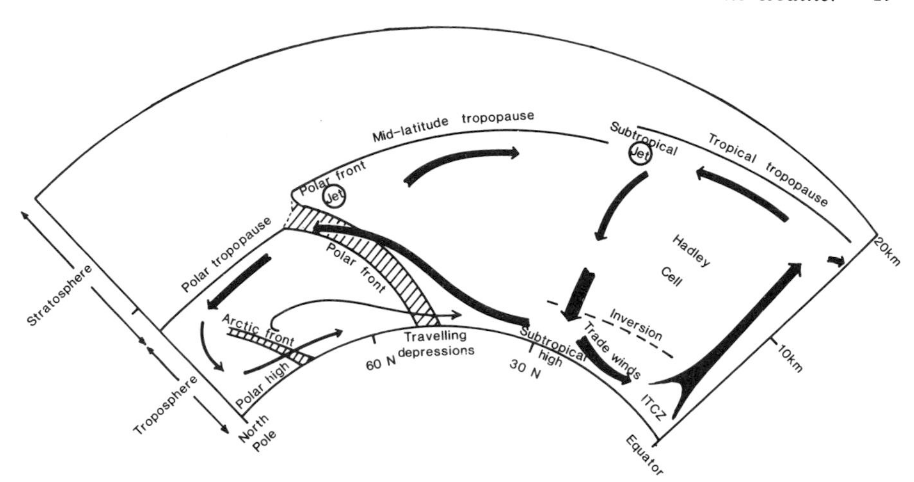

Fig. 1 Simplified cross-section of northern hemisphere in winter, showing vertical circulation in the troposphere. Jet streams blow from west to east. Arrows crossing fronts indicate latitudinal movement of front, not airflow through front.
(thick arrow) warm air (thin arrow) cold air

eastwards in both hemispheres by the earth's rotation. Some of this high level air creates a belt of westerly winds at about 30° latitude while the remainder cools by radiating heat into space, becomes denser on cooling, and sinks.

This subsiding air creates a zone of high pressure at the earth's surface in the subtropics while the ascending equatorial air results in a surface trough of low pressure in the tropics. Again to achieve a balance, air flows from high to low pressure (this time at low levels) with some of the subtropical air moving pole-wards to contribute towards further circulation at higher latitudes, while the remainder flows equator-wards as the Trade winds. The Trade winds from the two hemispheres converge in the equatorial trough to form the Inter-Tropical Convergence Zone (the ITCZ) within which is the convective uplift of tropical air described earlier. The ITCZ, often known as the doldrums, moves north and south with the sun in a seasonal movement – only slightly when over the oceans where water temperature changes little with the seasons, but markedly when over landmasses where heating is pronounced. The vertical circulation so described is known as the Hadley Cell, after the 18th century meteorologist who first proposed it (Fig. 1).

The circulation at higher latitudes is controlled by a similar high to low pressure flow. The radiation loss to space from the polar atmosphere exceeds the gain, and a net cooling takes place. As we have seen, temperature contrasts set up pressure differences and, together with the Coriolis effect, a strong west to east airflow is created in the upper troposphere of mid latitudes. The temperature contrast is greatest in winter, when no radiation at all is received at the pole, and this is mirrored by the strength of the airflow. The strongest winds are in a band known as the polar jet stream, which varies considerably in direction, velocity and position, owing to a wave-like motion in the tropospheric flow. These waves are responsible for the transfer of energy by large scale eddies between high and low latitudes. Their maintenance is assisted by other inequalities of temperature which arise from differential heating of oceans and landmasses, and obstructions to the westerly flow in the form of mountain ranges. For example, the Rocky Mountains rise to an altitude of 2 to 4 km, presenting a barrier which is aligned in a northsouth direction across the flow through over 47° of latitude. Their effect on the airflow is somewhat analagous to that of a rock projecting upwards from a stream bed.

Within the tropospheric waves, pole-ward and equator-ward movements of air exist in which cold polar air is carried to low latitudes, and warm air from subtropical high pressure zones is carried to high latitudes. The boundary between these airmasses is known as the polar front, and can be traced in winter more or less continuously round the earth. It is approximately parallel to, and beneath, the jet stream, meandering through the mid latitudes of each hemisphere. Other jets and fronts exist, but the polar front and its jet are the main features of the mid latitude circulation (Fig. 2). Henceforth in this section, I will consider only the northern hemisphere.

Because of variations in density, the less dense warm air overrides the cold air along the polar front, which slopes with altitude towards the cold side. As the contrasts of temperature and density are never precisely balanced, waves are generated on the sloping frontal surface. Energy is re-distributed in the vertical movements, and pressure falls at the earth's surface, most markedly near the wave tip (Fig. 3). There are certain areas over the planet which favour development of such waves, notably where topographical or oceanographical features enhance the airmass contrasts. A significant area with regard to European weather is the sharp transition in sea temperature at the edge of the Gulf Stream current off the eastern seaboard of North America.

The variation of pressure outward from the tip of a wave sets up a low level pressure gradient. This results in an inflow of air from high to low pressure in an attempt to compensate for the depletion of air at the centre due to upward motion. The Coriolis force deflects this converging air to the right, and it flows anticlockwise round the area of low pressure. Known as a depression, the resultant vortex is one of the major eddies in the atmosphere.

A depression has both vertical and horizontal circulation. The vertical circulation is caused by the lifting of the warm air by the cold air as the depression moves horizontally beneath the jet stream in the upper troposphere. In its early stages the depression is carried by the jet in a northeasterly direction in the pole-ward branch of the upper tropospheric waves described earlier. A deepening depression eventually distorts the upper airflow as its own horizontal circulation spreads upward through the troposphere, and ultimately it decelerates.

The warm airmass to the south of a waving polar front is known as the warm sector. That part of the front at the leading edge of the warm sector is the warm front, while at the rear edge of the sector (i.e. the forward boundary

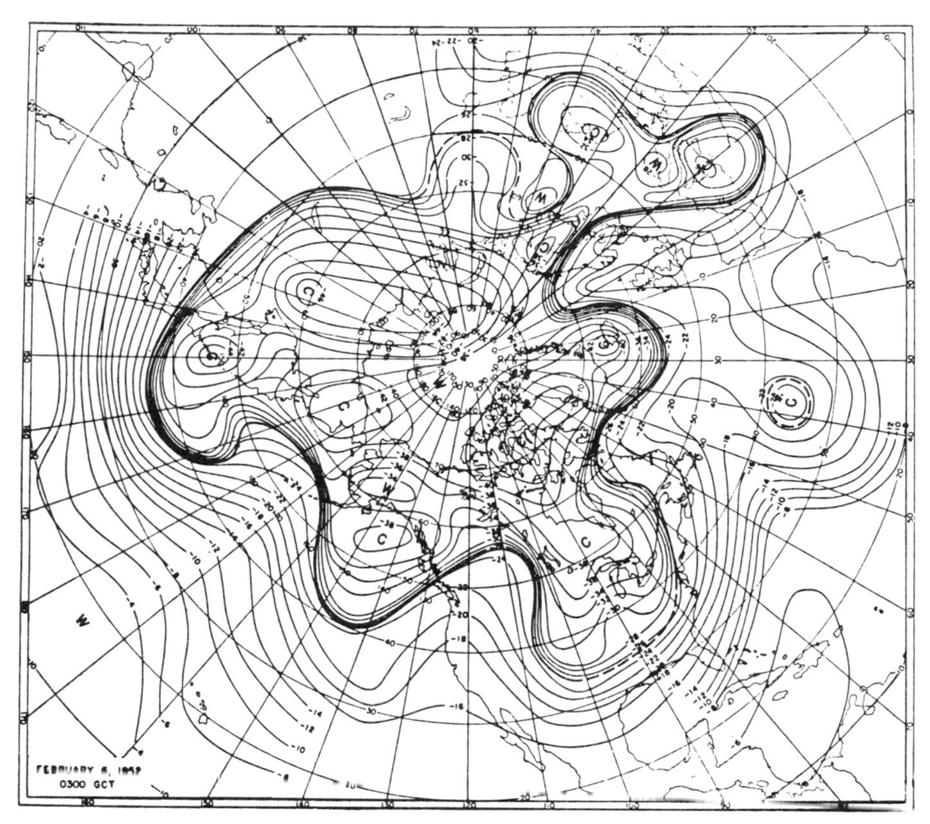

Fig. 2 Temperature distribution in the middle troposphere at the 500 mbar level in winter (6th February, 1952). Altitude approx. 5·4 km, temperatures in °C. The crowding of isotherms along the boundary of the polar front is exceptionally well-defined, meandering in a wave pattern round the northern hemisphere. It separates polar and tropical air, and is approximately contiguous with the polar front jet stream. Troughs of cold air and ridges of warm air indicate the latitudinal movement of cold and warm airmasses. Pools of cold and warm air north of the front are associated with the circulation of mature depressions. The heavy line marks the approximate southern limit of the polar air, including the frontal zone. Where the latter is weak, the line is dashed. The surface position of the sloping front is on the warm side of the 500 mbar position (from Bradbury and Palmén 1953).[18]

of the cold air) is the cold front. The warm front is carried east and then northeast in the depression's circulation, while the cold front moves south and then southeast and east behind the depression (Fig. 3). The cold air normally travels faster than the warm air, eventually lifting the latter from the earth's surface. As it does so, it overtakes the cold air which was originally ahead of the developing wave depression. The boundary between these two cold airmasses is called an occlusion, and although the airmasses originated from one, changes in character while circulating in the vortex finally make

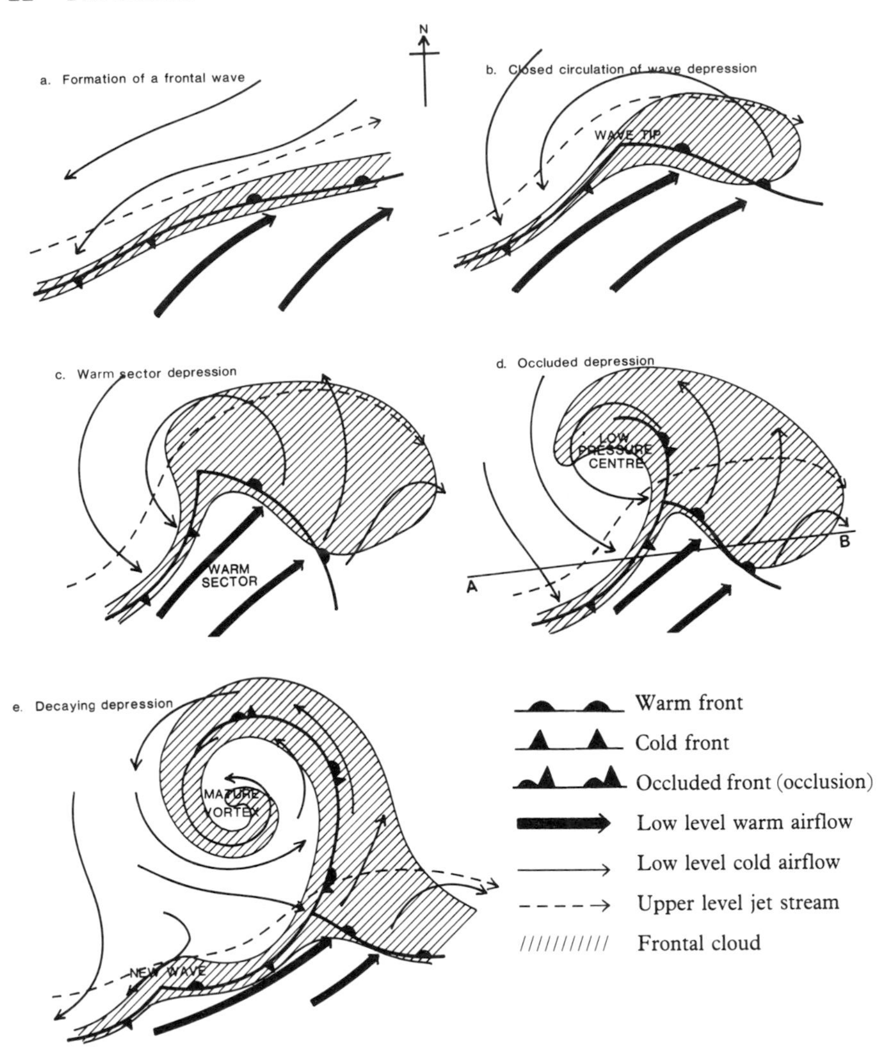

Fig. 3 Stages in the development of a depression. The fronts are at surface level.

them slightly different. The warm air, still above the occlusion, rises to high levels, cools and then sinks at higher latitudes to return south in due course as cold polar air.

During the formation of the depression, warm air is rising steadily over the cold air at the warm front, and is being undercut by the cold air at its rear edge. This rising air ascends into a region of lower atmospheric pressure. As it does so, it expands and cools. Vast amounts of water vapour are continually being evaporated from the earth's oceans, especially in the cloudless areas of the subtropical high pressure zones, and the warmer the air the more water vapour it can carry. This vapour is carried along with the airmasses and, when cooled on ascent, condenses to form cloud. On the polar front, where there are large masses of air rising and cooling, cloud is both abundant and deep.

The occlusion process heralds the beginning of the end of the depression's activity. It slows down, the front weakens, and eventually the pressure in the centre rises as the vertical circulation ceases. While the depression slows down, the frontal cloud revolves in the circulation. An old depression is characterised by spiral bands of cloud with the clearer polar air between the bands (Plate 1). These cloudy areas finally become confused as the airmasses mix and lose their identity. In the polar airmass behind the depression lies deep cold air while, further away, air descending from aloft forms an area of high pressure. Well to the south of an occluded depression, the polar front is still present, though relatively inactive. Further disturbances may arise and a new wave may form. If it deepens it will become the next depression, though some waves merely run along the front and temporarily enhance activity during their passage. Thus the process is repeated, with a continuous sequence of travelling depressions interspersed with transient areas of high pressure (Fig. 5).

The horizontal and vertical circulations of an area of high pressure are in a reverse direction to those in a depression. Owing to contraction as the descending air subsides gently into a region of higher pressure, the air warms and water droplets evaporate. The horizontal circulation is clockwise and, if complete, the area is known as an anticyclone. Anticyclones are noted for their quiet, often cloudless, weather as opposed to the windy and cloudy conditions associated with a depression. The polar anticyclones in the cold air between depressions are normally transient compared to those in the subtropical warm air, since they also move along under the upper air flow. However, there are semi-permanent anticyclones over continental interiors in winter, where very cold, dense and heavy air stagnates on the earth's surface.

The exchange of heat in the movement of airmasses between low and high latitudes is an indication of the continual attempt by the atmosphere to

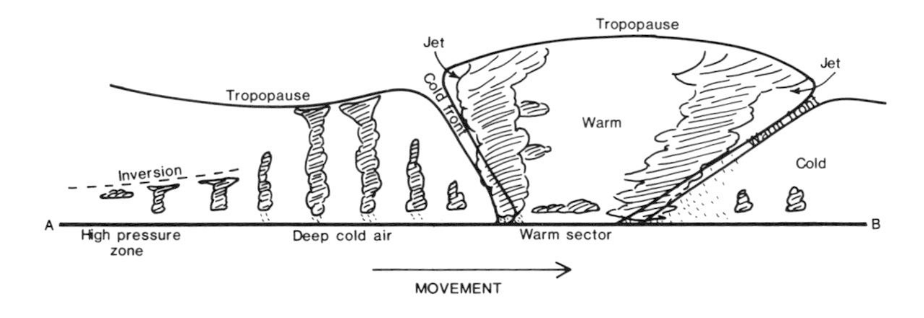

Fig. 4 Vertical cross-section of a frontal depression through the warm sector (A–B section of Fig. 3d).

balance temperatures over the globe. The most impressive direct evidence of this is the movement of air within those tropical storms which evolve into frontal depressions. The air in such a disturbance above the North Atlantic originates over West Africa. It is carried west in the Trade Wind zone towards the Caribbean Sea, turns northwest and then northeast along the eastern seaboard of the USA, to meet cold polar air in mid latitudes. The dying tropical storm is re-invigorated as the polar air lifts the warm tropical air, finally moving northeast as a frontal depression towards northwest Europe. Not a few of the first autumn storms in Britain originate in this manner, and may eventually carry their air of tropical origin as far as the arctic coast of Russia.

Airmasses owe their contrasting characters to their prolonged stay in one area prior to moving into the mid latitude circulation. Tropical air originates in subtropical anticyclones, whereas polar air is formed in cold regions of stationary polar anticyclones. Within these regions the airmasses are able to take up definite features from the underlying surface (e.g. temperature and moisture content) over a considerable period of time before moving gently to the edge of their high pressure system and into a more mobile regime where modifications take place.

Airmasses therefore have typical characteristics which depend on their track and place of origin. Tropical airmasses moving over the sea hold a great deal of moisture, with the result that much cloud is formed when the lower layers are cooled on traversing colder seas. This tropical maritime air approaches Britain in warm sectors. Above heated continental regions in summer, the air is often dry, and tropical continental air arrives in Britain from the south or east to give fine weather. Polar maritime air arrives behind cold fronts from the Atlantic, giving showery conditions over Britain, particularly in winter when the air is much cooler than the sea. Summer showers occur over land when convection is initiated by daytime solar heating. Polar continental air is dry and cold, and is a winter phenomenon. It brings fine frosty weather to Britain, though when approaching over the North Sea, shallow convection often gives dull cloudy weather in coastal areas. Modifications by land and sea surfaces therefore determine the type of weather that an airmass brings. Atypical features are not uncommon, however, since an airmass often undergoes many modifications during its lifetime.

From the description of a typical depression, it is possible to picture the sequence of weather as the frontal system reaches an observer. Referring to Figs. 3d and 4, it can be seen that to the southeast flank of a depression the warm front will be the first feature in evidence. The sloping boundary of the warm air first appears at high levels with wispy clouds of ice crystals, known as Cirrus. The wind is from a southerly point and strengthens as the front approaches, accompanied by falling pressure. As the warm air boundary lowers, its layered cloud deepens and rain begins to fall. The passage of the front at the surface brings a halt to the pressure fall and a rise in temperature

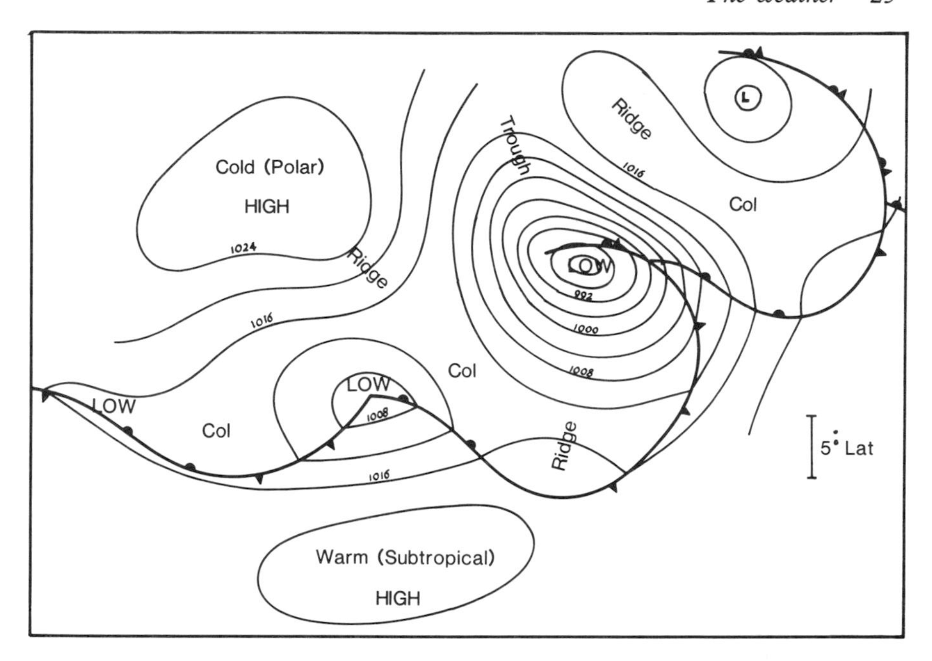

Fig. 5 A family of depressions, illustrating isobaric patterns.

in the warm air. The rain begins to die out. In the warm sector, the wind veers to the southwest and there may be breaks in the rather shallow low cloud. The cold front – sloping backwards – comes relatively suddenly, and with a period of quite heavy rain from thick and often convective cloud, perhaps accompanied by thunderstorms. Since the slope is steeper than that of a warm front, the cold front passes more rapidly and there is often a sharp clearance to cloudless skies. The temperature drops in the polar air, and the wind veers towards northwest. The pressure, which has remained fairly steady in the warm sector, begins to rise. For a time, the cold airmass may produce showers, but these tend to die out later under the influence of the subsiding air of the anticyclone behind the depression. If already occluded, the depression may bring only one front, and one period of cloud and rain. Wind strength in the vicinity of a depression depends largely upon its depth or rate of deepening, and a shallow depression is often accompanied by light winds.

The precise detail of the sequence at any point varies infinitely with the track, speed, depth and age of the depression, and also with the observer's position relative to the depression and the distribution of land and sea about him. Additionally, a depression often undergoes rapid modification once it reaches a land mass, and depressions over continental regions such as North America differ in detail from those generated over the oceans – chiefly due to variations in moisture availability. Other 'lows' may be formed without organised frontal systems, particularly in areas of convection such as summer thunderstorms. The sequence described is one fairly typical of a mobile Atlantic depression passing north of an observer in western Europe, and on many occasions, the signs can be read with sufficient accuracy by an observant person to predict short-term events.

Although disturbances on the surface are frequently transient, the large scale waves in the mid and upper troposphere are more slow-moving. Their amplitude and wavelength vary; sometimes they become more or less stationary with the result that changes on the surface are slow. This normally happens when the amplitude becomes sufficiently large to cut off the more typical west to east flow over a considerable latitudinal distance, and is known as a blocking situation. Blocking tends to inhibit the movement of surface low pressure systems, giving rise to prolonged periods of a similar type of weather over a particular area. The immediate causes of such periods are usually clear, but the ultimate cause of circulation anomalies is invariably obscure. Many possible inter-related factors may be involved, but one of the more significant may be anomalies in sea temperature on a large scale. Indeed, global weather patterns are now believed to be modified in a fundamental manner by ocean surface temperatures, both in the short and long term. Areas of abnormally cool or warm seas act as reservoirs of cold or heat, since their temperature changes only slowly over a long period.

PRESSURE AND AIR MOVEMENT

The pressure indicated by a barometer is a measure of the weight of the atmosphere above the instrument, and pressure decreases with altitude. The rate at which it decreases depends on the density and temperature of the air. The pressure at any given altitude is generally lower in a cold airmass than in a warm one and, as already described, this results in a pressure gradient which is the main feature of the global circulation.

Although the mean pressure at sea level over the earth is approximately 1013 millibars (mbar), the actual figure varies considerably, attaining 1040 mbar in a well-developed anticyclone, and 950 mbar in a vigorous winter depression. Values exceeding these are not uncommon, and pressures are particularly low in tropical storms.

Much lower values are found in mountainous areas. For example, the mean pressures at the summits of Ben Nevis, Mont Blanc and Mount Everest are about 860 mbar, 550 mbar and 320 mbar respectively, and the density and oxygen concentration (affecting high-flying birds) at these levels are, respectively, 90%, 60% and 40% of the values at sea level. A one millibar change at sea level is equivalent to 8 metres in altitude.

In meteorological data, barometric pressures are necessarily reduced to mean sea level to achieve conformity. They are essentially measurements of static pressure i.e. the pressure of the atmosphere alone, shielded from the influence of the horizontal airflow around the instrument. The pressure of the airflow on an object is called the dynamic pressure, and is incidentally one of the bases for measuring the speed of an aircraft in flight. Dynamic pressure can cause almost instantaneous fluctuations on an unshielded barometer of up to 5 mbar in a very strong gusty wind. In contrast, static pressure changes are small, exceptionally reaching 3 to 4 mbar per hour in the vicinity of a deep depression.

One of the main features on a weather map (which meteorologists call a synoptic chart) is the distribution of barometric pressure shown by isobars, lines joining points of equal pressure at sea level (Table 1; see also Fig. 5). Due to the earth's rotation, as already described, the airflow resulting from the pressure gradient moves anticlockwise round a depression, and thus follows the trend of the isobars. A set of isobars close together indicates a steep pressure gradient (e.g. Fig. 41) and, consequently, strong winds. The converse situation indicates light winds. The curvature of the isobars is important. If they are concave towards low pressure, the curvature is called cyclonic. If concave towards high pressure, then the curvature is anticyclonic. The former, called a trough of low pressure when the curvature is marked, implies general ascent of air, and there is frequently a marked kink in isobars where they cross fronts. The latter, known as a ridge of high pressure, implies descent or subsidence of air. The weather and cloud associated with these isobaric patterns tend towards the type described previously for depressions and anticyclones. A further pressure pattern is the col – an area between two depressions in which the horizontal pressure gradient is very weak, characterised on the map by a dearth of isobars.

Owing to the vigorous circulation of a depression, the winds in its vicinity are much stronger than in an anticyclone. Below about 1 km altitude (within the 'boundary' layer) friction from the underlying surface deflects the wind slightly, with a component towards low pressure. The deflection is greater over land than over sea. Thus surface winds have a component inwards towards a depression, and outwards from an anticyclone.

In vigorous and deep depressions, and also in tropical storms, surface winds may reach 30 to 40 m/s, with still stronger gusts, but they are normally less than 10 m/s in areas of high pressure. Extremely strong winds are often referred to as hurricanes, but it should be emphasised that this term is correctly applied only to tropical storms with winds exceeding 33 m/s. Winds of that strength elsewhere are only of hurricane *force* (Table 2). In the jet streams of the upper troposphere, winds exceed 30 m/s, and speeds of 75 to 100 m/s are not uncommon. Velocities of this high order can also be reached on the surface in special circumstances. Tornadoes generate such winds very locally, and are vortices of rapidly moving air sucked up into the vast thunderclouds which form in summer in heated continental interiors, notably in North America. Violent winds can be found at the edges of icecaps in winter, particularly where depressions are associated with very dense cold air pouring rapidly over the coastal regions of Greenland and Antarctica.

Vertical air movements are generally one to three orders below those of horizontal airflow, especially in the mass ascent of warm frontal zones, where vertical speeds of 5 to 10 cm/s are found. Some quite substantial values are generated, however, during the convection process. Air may rise to the tropopause under some conditions, but only a few hundred metres in others. Whenever there is ascent, there must be a compensating descent of air at some point in the vicinity. Convection begins when cool air is heated by a warm surface, be it land or sea. Rising, buoyant bubbles of air are formed, and are known as thermals. They reveal themselves as cloud if their water vapour content is sufficient to condense when cooled. Cumulus cloud results at this stage, changing into Cumulonimbus if the thermals rise to heights at which the water droplets freeze into ice crystals. If thermals are shallow, or form over arid areas, no cloud appears. Further details of convection can be found in Chapters 2 and 4.

In small Cumulus clouds, and in dry thermals, vertical speeds in up-draughts may reach 1 to 5 m/s, but in large Cumulonimbus clouds, speeds of 3 to 30 m/s may occur, with rather weaker downdraughts. In very severe thunderstorms, such as those in continental interiors in late summer (but uncommon in Britain) vertical velocities in updraughts can exceptionally reach 100 m/s, with downdraughts in the same cloud in the order of 40 m/s. Forest and grass fires, and even burning stubble, produce their own thermals – the intense heat often resulting in localised upcurrents of 10 m/s or more.

Moist thermals (thermals within cloud) can reach altitudes of up to 13 km in Britain and 20 km in tropical conditions. Dry (or 'blue') thermals may only rise to a maximum of 5 to 6 km, and then only in hot desert regions. Differences in the maximum height attained by cloud tops are normally due to variations in the height of the tropopause, which can be as low as 7 km in polar regions, but as high as 18 km in the tropics.

Vertical movement of air is also found where an airflow meets an obstacle in its horizontal path. This might be as small as a sand dune or an ocean wave, or as massive as a high mountain range. Over high ground, the enforced upward motion that air must undergo to cross the barrier is known as orographic flow. Under certain temperature and wind conditions, further upward motion may be found downwind of the barrier in standing waves, and vertical velocities in these waves can be in the order of 1 to 10 m/s or even more. The crests of the waves are often capped by cloud (wave cloud), and waves and clouds are each relatively stationary. Further details will be found in Chapter 2.

TEMPERATURE

The temperature of the atmosphere ultimately depends on the amount of radiation received from the sun, either directly or indirectly, via the earth's

Hills obscured by rain clouds

surface. Since the atmosphere is mostly transparent to this radiation, with important exceptions such as water vapour and stratospheric ozone, it is the earth's surface that is heated initially, and various processes subsequently carry this heat upwards. One process – that of conduction – is very slow, affecting only the lowest few hundred metres of the atmosphere at most. Cloud layers reflect solar radiation from their tops, and earth radiation downwards from their bases, thus reducing daytime heating and nocturnal cooling of the earth surface. However, the principle processes involved in heat exchange are those of vertical convection and the slant-wise ascent in frontal systems.

The net result of these exchanges is a vertical temperature structure in the troposphere in which temperature falls with altitude, but not at a constant rate. The rate of decrease, called the lapse rate, depends to a large extent upon the amount of water present, both in vapour and liquid form, and the vertical air currents. I have already shown that air cools when rising, and warms when sinking. These changes in temperature take place with no outside influence, and are termed adiabatic. Rising air moves into regions of lower pressure, and thus expands, causing its temperature to fall adiabatically. It cools at a rate of about 1°C per 100 m, until its water vapour condenses and it becomes saturated. At this point, heat released during the condensation process reduces the lapse rate. Conversely, sinking air moves into a region of higher pressure and so contracts, causing its internal temperature to rise adiabatically.

The vertical temperature lapse rate of any particular mass of air determines whether or not a warmed bubble of air can become buoyant. If the lapse rate of the surrounding air is high i.e. the temperature falls quickly with height,

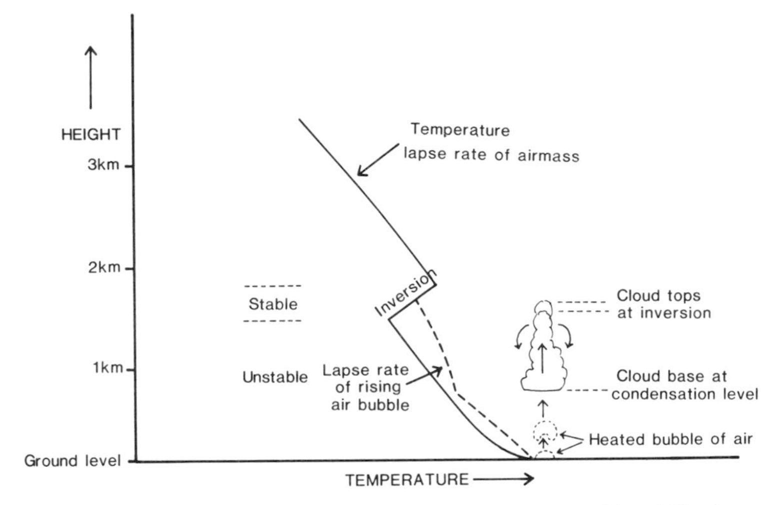

Fig. 6 Lapse rate of temperature associated with the formation of instability (convective) cloud.

then the bubble will remain warmer and less dense than its surroundings, and rise. The airmass is unstable. If the lapse rate is small or negative, the buoyancy of warmed air is lost, and the airmass is stable. Lapse rates in an airmass can change markedly with height, and it is these changes that contribute towards variations in cloud structure. A negative lapse rate, in which temperature rises with height, is known as an inversion, and is an extremely important feature – placing as it does a lid upon vertical air movement (Figs 6 and 7).

There are three situations in which inversions are commonly found. In the vicinity of a front, where warm air overlays cold air, the vertical temperature structure is such that in the sloping, mixing zone of the two airmasses, the lapse rate may become small or negative for a short vertical distance. Secondly, when the earth's surface radiates heat into space on a cloudless night, the cooling is conducted slowly upwards into the lowest layers of the atmosphere, and the air near the ground becomes colder than that immediately above it. This normally occurs over land, but the same effect is produced if warm air is cooled by flowing over a colder sea. The third case arises in anticyclonic conditions, where air from aloft warms on contraction when subsiding. At the lower boundary of this subsiding air, an inversion is formed, with cooler air beneath. The two latter cases are often combined, and in both the temperature may rise by 10°C or more in a short vertical distance. These latter inversions also inhibit the upward transport of water vapour, so that the air above the inversion is dry, while persistent cloud may exist in the moister air beneath.

Air temperatures close to the earth's surface show a wide range of values – as high as 40°C in an arid subtropical continental summer, and as low as

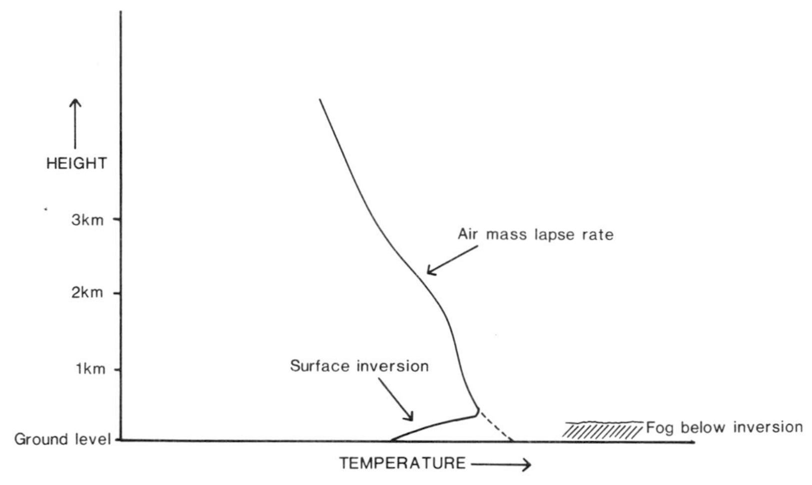

Fig. 7 Lapse rate of temperature associated with the formation of a low level inversion and fog. Fog disperses when surface temperature rises to modify lapse rate to that shown by dotted line.

−60°C in winter polar continental air. At an altitude of 5 km it may be −40°C in an arctic airmass, or 5°C in one of tropical origin. The temperature of the sea, and also that of large lakes, exerts a powerful influence upon the air temperature over the water and adjacent land masses. Temperature changes in the surface layers of the open sea are very much smaller than those over land, due to turbulent mixing and downward transport of heat. Diurnal variations are usually less than 1°C. This results in the general rule that the sea is warmer than the land at night and in winter, but cooler by day and in summer.

ATMOSPHERIC OBSCURITY

Visibility in the atmosphere can be reduced by suspended particles of dust or water droplets. The former is common in dry weather, especially in sandy arid regions where particles are raised by the wind, often to considerable heights and over long distances. Smoke haze is a frequent cause of poor visibility in areas heavily-populated by man, especially in winter, and also near forest and grass fires. Water droplets, however, are the primary agent for reducing visibility, and fog, mist and cloud are all manifestations of water droplet concentrations. Sea spray may reduce visibility when seas are rough.

Fog is essentially cloud on the earth's surface, and in meteorological parlance includes all occasions when the visibility is less than 1 km – poor visibility above this figure being known as mist. It is particularly frequent in cloudier hilly regions, and visibilities in cloud and hill fog may be reduced to a few tens of metres. Fog, as opposed to hill fog, forms when moist air near the earth's surface is cooled so that its water vapour condenses. This can happen either during radiational cooling on a clear night, or by contact with a colder surface such as the sea or snow. An inversion at low levels accompanies the fog, preventing an upward diffusion of water droplets (Fig. 7).

Cloud and hill fog are similarly formed, but cooling of the moist air is achieved mainly by upward motion. The greater the water vapour content, the more rapid the formation. Cooling arises by convection, in which cloud formation is vigorous and its extent in the vertical is frequently greater than in the horizontal; or during slow ascent, when clouds form in layers and the horizontal extent greatly exceeds that in the vertical. Slow ascent is found in frontal situations, or in turbulent flow where eddies are sufficient to mix the moist air and form shallow layer clouds.

Water droplets can exist at very low temperatures in a supercooled state, and in the atmosphere most clouds at temperatures above −15°C consist of water. Below this temperature the ice crystal content increases until at −40°C clouds consist entirely of ice.

PRECIPITATION

Precipitation – most commonly in the form of rain or snow – is typically initiated in deeper cloud by the freezing of tiny water droplets which have formed on a solid nucleus. These nuclei generally result from sea salt carried upwards from the oceans, but can also originate from desert dust. Where water droplets and ice crystals co-exist at high levels, the droplets are attracted and frozen to the crystals, which then grow in size. In their subsequent gravitational fall through the cloud, crystals collide with further droplets or crystals and eventually a snowflake – a conglomeration of ice crystals – is formed. Snowflakes begin to melt when reaching temperatures above 0°C, and the form of precipitation at the earth's surface depends upon the level above the ground at which this temperature lies. It is known as the freezing level, although a more appropriate term is the melting level. Precipitation can also be initiated in water droplet cloud when air movement is sufficient to cause droplets to grow by coalescence, and in low cloud often results in small-sized, drizzle droplets.

The intensity of precipitation depends upon several factors, the most important of which are the depth of cloud and strength of vertical air currents. In convective cloud, the falling precipitation must overcome strong upcurrents before reaching the ground. The stronger the updraughts, the longer a particular raindrop or snowflake will remain within the cloud, and the larger it will grow. The ultimate size is limited, since at a certain point it breaks into smaller particles. Raindrops in vigorous convective cloud may be frozen by being carried repeatedly upwards into the colder regions of the cloud, eventually forming hailstones – their size dependent on the strength of the upcurrents. The cloud depth influences intensity in a rather similar way

1: Infra-red satellite image, 1941 GMT, 16th September 1978. The large vortex north of Shetland (Low B, Fig. 22) originated as hurricane 'Flossie'. The infra-red image shows shades representing temperature, so that the coldest (highest) cloud is white, while warmer (lower) cloud is darker. Land is discernible under the clear skies, e.g. east Greenland, west Iceland, parts of northern Ireland and eastern Scotland, southeast England, the Netherlands, and parts of the Baltic and northern Scandinavia. The band of white cloud across England is contiguous with the jet stream and is the broken upper cloud of the weakening cold front. Wave clouds appear as ripples to the lee of the Welsh mountains. Spiral bands of clearer polar air have been swept into the vortex. The vast deck of upper cloud ahead of the warm front and on the occlusion covers much of Scandinavia and the Norwegian Sea. A clear area exists over the sea to the south of Iceland, where the unstable northwest airstream still has to initiate convective cloud as it streams from the cold land across a progressively warmer sea. The beginnings of convective shower clouds can be seen, becoming deeper cellular features towards Scotland. A vortex visible in the Barents Sea no longer has deep cloud and is in its last stages, having been a deep Atlantic depression six days previously. Low B itself finally died out a week later in the same area. *(Dundee University)*.

2: (a) Shallow Cumulus in 'streets'. The early stages in the formation of convective cloud. Thermals have been initiated in which the water vapour has condensed to form cloud at a uniform level. The cloud elements are drifting downwind of their thermal sources. Further vertical growth of convective cloud is determined by the temperature structure in the surrounding air. *(Norman Elkins)*.

(b) Mountain wave clouds. Cloud is formed by the condensation of water vapour in the rising air of a lee wave. Evaporation takes place as the air descends beyond the wave crest. However, the moisture content of the air may be too low for any cloud to be formed, or so high that a continuous cloud layer exists within the wave system. *(Norman Elkins)*.

3: Glazed ice. A winter hazard formed when rain falls through air at sub-zero temperatures and freezes onto objects. Persistent glaze results in large scale starvation of many species. In the photographs clear ice deposits completely cover branches, twigs and foliage of (a) deciduous trees, and (b) evergreens. *(J. Dudley-Davies)*

4: (a) Redwings feeding in snow. Large numbers of birds survive severe weather on food put out by man. The Redwing on the left is typically aggressive – a strategy assisting survival when many individuals must feed on restricted food supplies. *(Dennis Green/Aquila)*.

(b) Stock Dove dead in snow. Despite man's activities many birds die of starvation in periods of frost and snow. *(Dennis Green)*.

5: (a) Whooper Swans on ice. Small bodies of water freeze over in prolonged frosty weather, particularly sheltered waters, driving waterfowl onto large lakes and coastal waters where huge congregations may occur. *(John Edelsten).*

(b) Estuarine ice. Mixed river and sea water of low salinity on estuarine flats freezes in cold weather, preventing waders from feeding until the next incoming tide melts the ice. In the photograph the tide is ebbing, leaving deposits of ice at high water mark, formed during the night when ground temperatures fell to −17°C. *(Norman Elkins).*

6: (a) Lapwings. Inability to feed in frost and snow may result in long-distance movement, particularly of gregarious species which flock in open country, such as Lapwings. A less severe environment is sought, followed by a mass return when the weather improves. The birds in the photograph are insulating themselves against the cold by fluffing out their plumage. *(Pamela Harrison).*

(b) Grey Heron. A species that undergoes considerable starvation in severe winters in Britain when wetlands freeze over. *(D. A. Smith/Aquila).*

7: Windthrow of trees. In January 1953, severe gales felled many trees in Scotland, changing mature woodland habitat to scrubland: (a) 150-year old Beech trees, (b) 50-year old Scots Pine plantation. *(Forestry Commission)*.

8: (a) Rookery. Along with other treetop nesting species such as the Grey Heron and Osprey, Rooks lose habitat and nest sites when trees are felled by gales. Strong winds may also destroy nests. *(Pamela Harrison)*.
(b) Whitethroat. The worst affected of several European summer visitors which winter in the Sahel savannah of Africa, where a drought in the early 1970s contributed towards degradation of habitat. *(A. W. Cundall/Aquila)*.

9: Cumulus and Cumulonimbus. The distant Cumulonimbus cloud has a fibrous ice-crystal upper part, with a shower falling from the horizontal base. The nearer Cumulus cloud has a typical, cauliflower appearance with turrets building as strong upcurrents rise through the cloud. Soon the water droplets will freeze and another Cumulonimbus will be born. Both cloud types are associated with convergence zones in the atmosphere which concentrate flying insects. *(Norman Elkins)*.

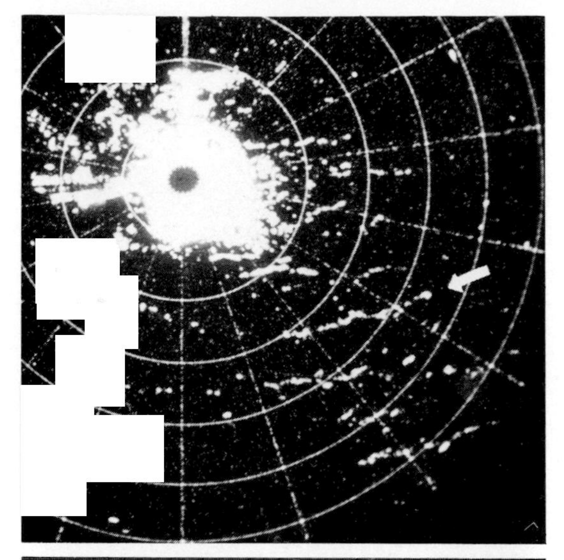

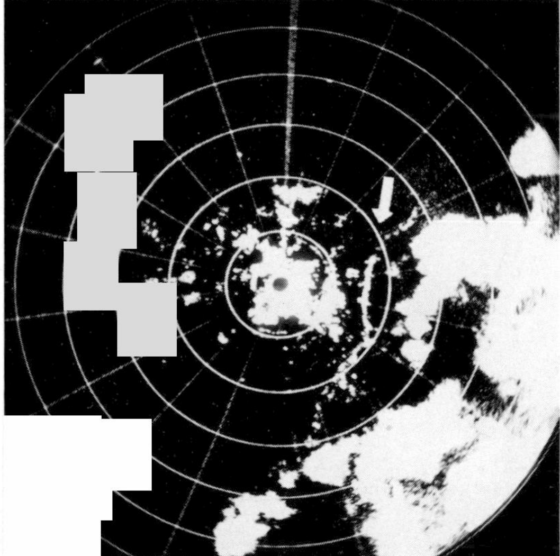

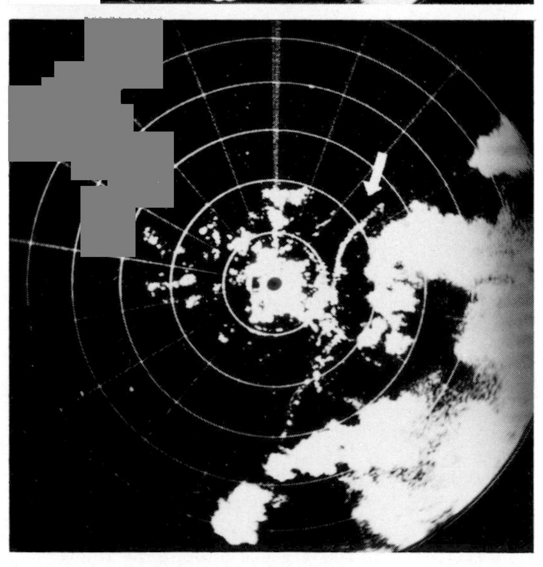

10: (a) Radar echoes of feeding Swifts in July. (i) Lines of 'angels' on radar (arrowed) at 1210 GMT. Range circles are at 8 km intervals. The lines lie several kilometres apart, roughly up-and-downwind, and are attributed to Swifts feeding in thermal 'streets' in fine weather. (ii), (iii) An intense arc of echo (arrowed), parallel to and several kilometres to the west of the edge of a Cumulonimbus mass: (ii) at 1703 GMT, and (iii) at 1711 GMT, showing the drift with the wind of the arc and precipitation echoes. The arc was attributed to Swifts feeding in upcurrents forming ahead of existing shower activity, probably in a gust front. (From Harper 1960[67]; *Crown Copyright*).

(b) Swift with throat pouch filled with food. The most aerial of insectivores, utilising concentrations of aerial plankton in the upcurrents of convergence zones, and exhibiting massive long-distance movements in summer to avoid food shortages in cold wet weather. *(Erik Hosking)*.

11: (a) House Martin building nest. The consecutive layers of drying mud can be clearly seen. In severe drought conditions birds have difficulty in obtaining sufficient suitable mud for nest building. *(W. S. Paton/Aquila)*.

(b) Sand Martin colony. Such colonies are normally transient in nature. Winter rains and summer cloudbursts wash away many existing sand cliffs, and also create new ones, often by erosion during spates. *(Pamela Harrison)*.

12: (a) Spring snowfall. Snow at high levels (in this instance in Scotland, at 640 m above sea level in early May) delays the nesting of many upland birds. *(Norman Elkins)*.

(b) Hen Ptarmigan on nest in snow. Although summer weather is not an important factor in the control of Ptarmigan populations in Scotland, the species nevertheless depends on snow melt for breeding sites and the growth of food plants. *(Dennis Green/Aquila)*.

13: (a) Great Crested Grebe on nest. For grebes and divers, water levels are critical in determining both the timing of breeding and its success. The Great Crested Grebes add material to floating nests as water levels rise, but flooding is a significant hazard for water birds with fixed nests. *(Pamela Harrison).*

(b) Reed Warblers at nest. The tall fragile vegetation favoured by the majority of breeding Reed Warblers is prone to damage or even destruction by summer storms. *(Kevin Carlson).*

14: (a) Eider ducks with creche of ducklings. Although ducklings are adapted for an aquatic life, low water-temperatures and cold wet weather pose problems in thermoregulation. More brooding may be afforded but some sea ducks lose many young in such conditions. *(R. W. Kennedy/RSPB)*.

(b) Sea caves on the Rock of Gibraltar. The caves provide a sheltered roost for several passerine species in winter, notably up to 3000 Crag Martins, which forage daily in nearby Spain. Eddies in the airflow ensure that, even in onshore winds, the cave interiors rarely lose their shelter value. *(Norman Elkins)*.

15: (a) Roosting Long-tailed Tits. To reduce heat loss in cold weather, some small passerines such as these Long-tailed Tits form compact clumps when roosting. *(H. Löhrl)*.

(b) Little Bittern panting. In high temperatures, birds lose heat by respiratory evaporation and by allowing the airflow to cool the skin beneath raised feathers. *(Kevin Carlson)*.

16: (a) European migrants are not infrequently blown out to sea and, probably, few survive unless the coast is near. (i) Exhausted Swallows on a ship 500 km west of Cape Finisterre in April. In stormy weather, low flying Swallows are occasionally engulfed by waves. (ii) A Chiffchaff on board a ship 250 km west of Ireland in November. *(T. D. Rogers).*

(b) Meadow Pipit and White Wagtail, off-course Icelandic migrants taking refuge on a weather ship in the Denmark Strait in spring. Such birds are frequent visitors to ships, particularly when meeting adverse weather. *(T. D. Rogers).*

in that the precipitation will remain within the cloud for a longer period if the cloud is deep, and increase in abundance while falling. Active and deep frontal cloud may also contain convective cells in which depth and upcurrents each enhance the precipitation, and this is especially true in the vicinity of cold fronts. Precipitation is increased on all occasions where cloud is lifted orographically.

In conditions of temperature inversion – in particular ahead of warm fronts in winter – rain can fall from warm air into a layer of air at subzero temperatures. If this layer is at the earth's surface, the rain freezes on impact, covering all objects with a clear deposit of glazed ice.

SUMMARY

I have attempted to explain briefly and simply the broad features of the atmospheric circulation; from the hemispherical wave-like motion of the mid latitude airflow and the vertical circulations of equatorial and mid latitude regions, through the sequence of weather in depressions and anticyclones, to the small scale (but still important) circulations associated with convection. More detailed descriptions of temperature structure, pressure distribution, air movements, visibility and the formation of precipitation follow, and emphasis placed upon the interdependence of each process upon the others. This inter-relationship is of utmost importance, since no single meteorological parameter is of significance in isolation. Weather is the end product of a whole range of complex physical processes.

CHAPTER TWO

Flight

Since birds are highly adapted for flight – more so than any other form of organism – it is appropriate that we should first investigate the effect of weather processes upon this feature. Among other organisms, only the insects have a large-scale flight capability, although in most cases they exert little positive control upon their movements.

Of the 9,000 or so species of living birds, about forty have lost the ability to fly and have adapted to other modes of travel such as swimming or running. Flying birds have a wide range of adaptations to withstand most of the atmospheric conditions that they experience. They have evolved degrees of flight which range from a life spent almost entirely on the wing, as in swifts and some seabirds, to a life spent on the ground where wings are infrequently used, as with some game birds. Those species which use airspace extensively are influenced the most by atmospheric conditions, and their flight behaviour is often controlled to a large extent by air movements. One finds that they have adapted fully to exploit this niche successfully.

I showed in Chapter One that the continually changing atmosphere exhibits movements of varying magnitudes. Air flow operates in vertical and in horizontal directions, and it is the vertical movement within the lower atmosphere that is of major importance to the more aerial of bird species. Flight imposes a constant drain on energy resources, and it is clear that the fuel used must at least be compensated by that gained from the intake of food. During breeding, moult or migration, energy is used at greater rates than normal and the intake must increase accordingly. The size of large birds determines the load they can carry in terms of weight, and the larger the bird the more severely limited is its active flight in altitude and range.[134] Thus any way in which a bird using an aerial habitat can reduce energy loss must be of immense benefit.

Many species have therefore evolved in behaviour and physical attributes to exploit those air currents which can minimise the effects of gravity, i.e. currents which have a vertical component upwards from the earth's surface. Vertical airflow is found in many degrees of form and magnitude, but can be placed into three basic categories: convection, orography and turbulence.

CONVECTION

Convection produces upcurrents of varying intensity, commonly known as thermals. These are used by a variety of species, and provide lift which enables a bird to reduce its rate of sink and so maintain height or gain a greater altitude than would otherwise be possible, but with a minimal expenditure of energy. The method of using these rising air currents is known as static soaring, and energy saving is such that several minutes of soaring are preferable to a few seconds of wing flapping. C. J. Pennycuick[135]

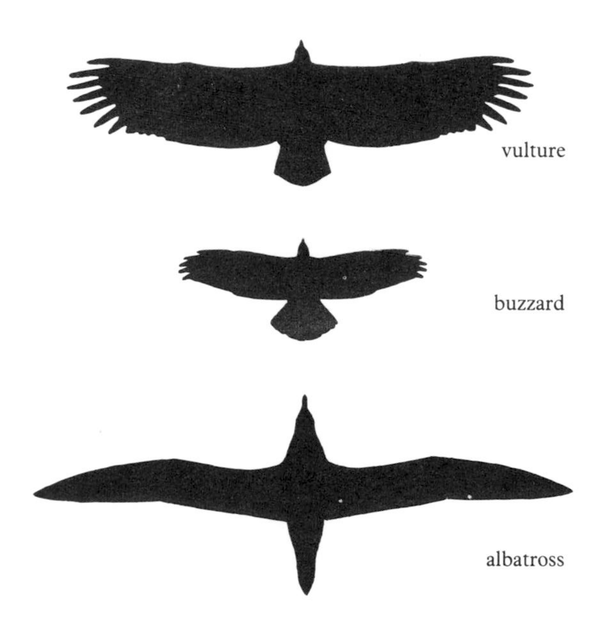

Fig. 8 Wing forms of soaring birds.

calculated that a large bird could reduce its fuel consumption by a factor of 23 by soaring rather than flapping.

Static-soaring birds typically have long, broad wings (see Fig. 8). This adaptation has reached the ultimate in the massive wing forms of vultures and condors, which have a low wing loading (weight per unit area) and use vertical air currents to the utmost. Their wings are deeply cambered, with large slots between the tips of the primary feathers, giving such birds the ability to fly at a low speed with a low rate of sink. Upcurrents enable them to soar in level flight or to climb in altitude during sustained soaring. Because they seldom use flapping flight, the flight muscles of the larger species are poorly developed and their habitat is confined to areas where upcurrents, of one type or another, are of more or less daily occurrence throughout the year. Many other raptors have evolved wings of similar shape, and broad-winged species such as the Golden Eagle and the Buzzard, with outstretched wings and 'fingered' primaries, are relatively familiar in northern Britain (Fig. 8). Other large birds which soar frequently include gannets, cormorants, pelicans, frigate-birds, herons, storks and ibises, and there are yet others, such as gulls, which use thermals when the opportunity arises. The larger gulls are not specialised fliers, but can exhibit a wide range of flying skills.

The origin of a thermal lies in a bubble of air which is warmed by a source on the earth's surface. Factors influencing the ability of a land surface to produce a thermal include the angle of the sun's incident rays, the dampness, moisture content and reflectivity of the soil, and the speed of the airflow over the surface. Since vegetation uses much of the incoming heat in transpiration, forested areas produce few and relatively weak thermals, while dry expanses of bare sand or soil normally make the most efficient thermal sources. Only a slight temperature rise is often sufficient to initiate a thermal, although in most cases at least a few degrees rise is necessary to sustain it.

Over land, then, thermals are initiated as soon as the sun raises the temperature at the surface, and provided that the solar radiation is of sufficient strength and is not inhibited by cloud, thermals will continue to be formed until cooling begins later in the day. As long as the temperature of the air bubble is higher than that of the surrounding air, it will remain less dense, and therefore buoyant, and will rise. It is the difference in these temperatures, not the temperature level, that is significant, and a high temperature is not necessarily a prerequisite for thermal formation. As the bubble rises, it cools adiabatically (see Chapter 1), frequently reaching a level at which its water vapour content condenses to form cloud (Plate 2a). The condensation level depends on the initial moisture content, so that the base of the cloud will be higher when the air is drier. This means that in arid areas only dry thermals are formed, while in maritime climates there is normally enough moisture available for all larger thermals to be capped with cloud (Fig. 9). A rising air bubble is, of course, not entirely isolated from its surroundings, and some mixing is continually occurring around its periphery. Mixing tends to lower the bubble's temperature and reduce its buoyancy. Thus, many air

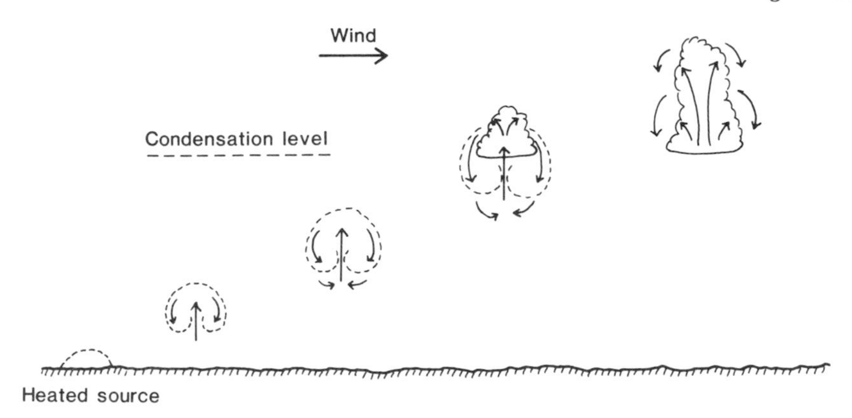

Fig. 9 Formation and development of a single thermal.

bubbles dissipate shortly after leaving the surface, but in vigorous convection, where the bubble is considerably warmer than its surroundings, it will reach a substantial height in favourable circumstances. Strictly speaking, a thermal consists of a series of rising air bubbles, and the greater their momentum due to initial buoyancy, the faster their rate of ascent and the higher their penetration before erosion by mixing. The width of the region of lift in a thermal may be 300–500 m at an altitude of 500–1,500 m, and greater at higher levels and in thunderclouds, which comprise several convection cells.

A brake on thermal activity arises when the moisture content of the air is high enough, and convection vigorous enough, to produce large masses of convective cloud. This effectively cuts off the solar radiation received at the ground, while the gentle descent of air outwith the cloud inhibits further thermal production, producing an earlier cessation of thermal activity than would normally occur.

Soaring birds use these expanding air bubbles to rise with little energy expenditure, provided that the vertical velocity of the air current is greater than their rate of sink. They respond quickly to changes in weather, and soar most towards midday when thermal activity is nearing its peak. Observations of raptors in the Coto Doñana, Spain,[69] showed that soaring activity increased with the morning rise in air temperature, and decreased from mid afternoon. This decline was assumed to be due partly to a weakening of thermals following the fall in temperatures which begin, usually, an hour or two after local noon. By this time, the buoyancy of the thermals may be reduced by mixing. This probably applies to most thermal activity over land, given that the overall meteorological conditions remain unchanged. With a strong wind, thermal production is somewhat weaker, but can still be abundant. Thermals move downwind, and those from the same source may form ‘streets’.

Soaring birds, particularly those on migration, may make use of such 'streets'. Birds can profit from convection by choosing a flight path which lengthens the time spent in rising air. This can be undertaken by flying over a thermal source whilst travelling in the preferred direction, then soaring slowly and gaining height before gliding rapidly through the adjacent downcurrents to a new thermal source.

Buzzards on Speyside in Scotland are most active in thermals from mid-morning to mid afternoon, with temperatures near the ground at or above normal, with wind strength force 1 to 4 (up to 8 m/s), and with partial Cumulus cloud cover. After some initial flapping, which is invariably necessary until a thermal has been found, the birds are able to soar at heights between 20 m and 600 m above ground level.[183] Raptors will rise to considerable heights if the upcurrents are deep, and there is evidence that Golden Eagles in Texas may reach 5,000 m. Sexual dimorphism in raptors may mean a difference in soaring ability; in some species the wing loading increases with weight, and the lighter males may be able to soar in weaker thermals than the heavier females.

Many of the largest soaring birds are so completely dependent upon thermal activity that their diurnal cycles are controlled almost exclusively by thermal formation. That they must seek thermal sources to assist take-off was illustrated by Betts[9] who described the flight of Adjutant Storks in southeast Asia. He watched them flap heavily from take-off towards the nearest thermal source of fallow paddy fields or hot expanse of asphalt, from which they could then rise effortlessly. Some subtropical raptors use thermals to rise from warm lowlands to high plateaux, in order to hunt for food. On the other hand, the onset of convective rain often coincides with downdraughts, forcing soaring birds to descend.

Pennycuick[135] observed soaring African birds from a motor glider, and noted that they used thermals and dust devils. The latter are distinctive thermals, formed over intensely heated surfaces, in which converging air at the base of the thermal accentuates whatever horizontal rotation may be present due to eddies in the airflow. The resultant vortex sucks up loose material – hence the name – and can be seen on a small scale at higher latitudes in the whirls of leaves, grass and dust on a hot breezy day. Pennycuick watched one vulture fly low in order to exploit dust devils in an area where thermals were poorly developed.

Even species such as Starlings use thermals on rare occasions. A flock of forty in the USA was seen soaring in a thermal with an Osprey. There was no reaction between the species but, since Starlings are ill-adapted to this mode of flight, they may have been simultaneously monitoring and duplicating the raptor's performance to enable the flock to travel to its roost with minimum energy loss.[24]

While thermals over land are initiated by solar heating of the ground, they will form readily when cool air flows over a warmer water surface. At high latitudes this is quite frequently the commonest mode of formation, and the only method in winter. The sea retains its heat for a much longer period than the land, and the range of temperature of the open sea surface is low – less than 1°C in 24 hours. In winter especially, but also at any season in outbreaks of polar air to low latitudes, the air is invariably cooler than the sea, and the resulting thermals may be abundant. With an onshore airflow in these circumstances, thermal formation ceases over a cold winter landscape, although existing thermals penetrate inland to some degree, usually being blocked by the first line of high ground. With an offshore airflow, formation will begin over the sea, but a clear area exists just offshore where moist thermals have not yet had time to form.

Few species use thermals over the sea – gulls being among the more frequent. Under calm conditions, gulls may be able to initiate their own thermals by circling in large groups and so extend their soaring flights to much higher levels. They can do this when the temperature difference between the cool air and the warm sea is between 3°C and 6°C. In calms, thermals will be readily initiated by circling, which imparts a certain rotation to the air. This is analagous, in an inverse manner, to the formation of desert dust devils. Sustained soaring of gulls has been observed at all levels of instability, even when thermals were only just detectable. With wind speeds above 2 m/s, thermals suitable for soaring are produced naturally, being most readily formed at wind speeds around 5 m/s – even in only marginally unstable air. Above 5 m/s, there is a direct relationship between the wind speed and minimum air-sea temperature difference for thermal soaring, so that the stronger the wind, the more unstable the air must be in order to produce suitable thermals. These are the thermals which are sufficiently organised to produce cloud systems. However, when wind speeds are high – above 12 m/s – soaring ceases, since the strong turbulence probably precludes sustained thermal soaring.[194]

Other seabirds, and land birds migrating over the sea, may also use thermals, although they are probably too unreliable to be of regular benefit to land birds. One group of seabirds noted for its soaring behaviour is the frigate-bird family of tropical regions. Here in the Trade wind zones, they are outstanding fliers, soaring extensively and often rising to 300 m. There is considerable low level convection in the Trade winds, and Cumulus clouds continually form and dissipate in the warm moist air. These clouds are often aligned in 'streets' along the wind, and over the Atlantic the convection is normally limited in depth to 2 to 3 km. It is within the Trade wind zone that late summer disturbances occasionally result in intense rotating convective tropical storms, and the Magnificent Frigatebird is said to be among the few birds able to ride out a hurricane in flight.[26]

In cold weather, upcurrents may be induced over land by man-made artefacts. In low temperatures, even below 0°C, heating from towns and cities – the well-known heat island effect – may produce limited thermals which are rapidly exploited by gulls. Flocks break off from direct flight to gain height by soaring in a localised thermal above a group of warm buildings, or a power station, before continuing on their journey. A Buzzard has even been recorded using a thermal produced by the gas flare of a hot air balloon on a cold windless day, during which it circled 20 m above the balloon, gaining about 70 to 75 m in height in 7 minutes.

OROGRAPHY

The second category of vertical air motion comprises those movements resulting from the enforced (orographic) ascent of air over high ground. Species which inhabit hilly terrain are able to use vertical components of air currents throughout the year, which are not dependent upon solar heating. Such soaring birds can range to higher latitudes and more maritime climates than those dependent upon thermals.

The vertical components of orographic airflow vary according to the strength of the flow, the distribution of wind velocity and direction with altitude, and the vertical temperature structure of the atmosphere. At its initial forced ascent over a hill, the airflow is generally smooth, but once beyond the crest it becomes more turbulent. Eddies form to the lee of the hill, and occasionally below the crest on the windward side, and if winds are strong the eddies cause violent turbulence. Under certain circumstances, standing waves may develop downwind, their wavelength and amplitude varying with the wind strength and temperature lapse rate. When the air is moist, cloud forms on the updraughts over the hill itself and in each lee wave, and may dissipate in each downdraught (Plate 2b). These lee waves are favoured areas for glider pilots, who find a line of rising air on the windward side of each wave crest (Fig. 10). Naturally, the airflow is much more complex in mountainous terrain where ridges of different heights cover a wide area and lie at various angles to the wind. In light winds there may be further subsidiary air movements, diurnal in character, superimposed upon the general flow. Daytime heating causes an upslope movement known as anabatic flow. On a fine day anabatic winds can reach 3–4 m/s but do not occur under total cloud cover, since temperature variations are suppressed. Convection may be superimposed on slope winds and lee waves, so that the latter especially are considerably weaker during the middle of a warm sunny day. Orographically rising air can, under certain conditions, become more buoyant as it rises, so that a hill crest may act as a thermal source without solar heating.

Thus the airflow over mountains is complex, depending on gravitational, frictional, thermal and orographic forces. Soaring birds must be aware of the ideal sites for rising air, reacting rapidly to changing situations. This is probably most valid in their home range, where the birds are sufficiently familiar with local topography to be aware of the location of thermal and orographic upcurrents, and the effects of changing winds. Raptors are especially adept at combining orographic and thermal soaring.

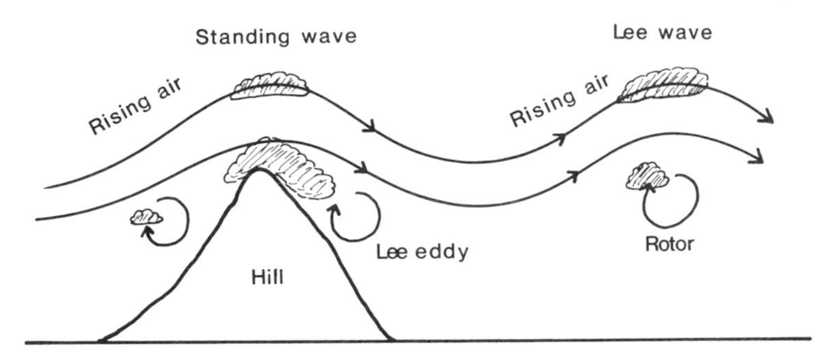

Fig. 10 Orographic airflow. The formation of rotors and strong eddies normally accompanies only the stronger wind flows.

Orographic soaring requires the same physical wing forms as thermal soaring. Many crag and mountain species have broad wings with large wing slots, and do not seem to be unduly hindered by very strong winds. Large raptors, and also choughs, are able to fly comfortably in winds of force 8 to 9 (17 to 24 m/s). Indeed, Leslie Brown[20] observed a Golden Eagle flying magnificently in a 50 m/s storm, seemingly unaffected by turbulence although it may have been in the smooth flow to windward of a hill crest. A measure of the ability of large raptors to gain height on rising air was given by Seton Gordon[60] who watched Golden Eagles rising 200 m up a vertical cliff in less than 30 seconds. Another bird glided downwards into a severe westerly gale towards the lee of a cliff, rocking in the downdraught, and then turned downwind at high speed. It returned later, flying rapidly along the cliff about halfway down in an easterly eddy.

Gannets soar in upcurrents over their colonies and, like other cliff species, will use lee eddies and standing waves. The latter are formed over isolated rocks and islands in the same manner as over hills. Use of these waves by Gannets, Fulmars and gulls has been observed to the lee of the Scottish gannetries of Ailsa Craig and the Bass Rock.[8] At the latter, up to three columns of soaring birds have been observed, gulls and Fulmars at the top (with lower wing loadings) and Gannets at the base. The columns related to the orographic ascent over the rock itself, and the rising air in two downstream waves. The birds circled in the upcurrents without flapping, finally gliding away, and Gannets approaching the rock used the ascending air to gain the peak with the minimum usage of energy. Lee waves and other waves due to wind shear can occur almost anywhere in the atmosphere, and it is theoretically possible for birds (particularly those on migration) to make use of them, although this is very difficult to confirm in the field. Waves and vortices may extend as far as 300 to 400 km downwind of peaks and islands, and birds probably recognise the cloud forms as indicators of rising air.

Even in the tropics orographic flow can be important. In Tanzania, Ruppell's Vultures breed in large numbers on cliffs exposed to the prevailing winds, which provide a continually rising airflow. They are therefore independent of thermals, early and late in the day, and are able to remain airborne for extended periods, a great advantage when seeking the migratory herds of Wildebeeste which may be up to 120 km from the cliffs. The Lammergeier often travels up to 40 km to foraging areas remote from its large mountainous territories, and is assisted by orographic currents and by thermal sources in adjacent plains and valleys.[26]

TURBULENCE

A third category of rising airflow is created in eddies wherever the airflow meets an obstacle. Gulls are expert at using eddies caused by the passage of a ship, and those present in the vicinity of buildings. Corvids and raptors deliberately use eddies at edges of woodland, and innumerable other examples could be cited. Eddies near the ground are also frequent in turbulence caused by friction when air moves quickly over a rough surface. Their rapidly changing vertical and horizontal components give rise to the familiar gustiness of a strong wind (Fig. 11). Eddies may cause difficulties in downwind flight, since there are risks both in high speed flight and in the sudden lulls that may reduce a bird's airspeed to below its stalling speed. Certainly, large birds like to take off into the wind under these circumstances. Flight is less hazardous at a higher altitude, away from the worst of the low level eddies. Conversely, against a strong wind, birds may fly very low, hugging the ground and following contours in line astern. By doing this, they gain a degree of shelter, and move in a layer of air in which the mean speed is reduced by friction. Thus there is a tendency to fly higher with a tail wind than with a headwind.

The flight behaviour of flocking species is often modified according to the strength of the wind. J. Brodie[19] described one particular flight formation of Starlings – a vast cloud-like column – and noted that the birds flew rapidly on windy evenings; but slow flight occurred when wind speed and temperature were each low. He suggested that in the former case the birds were using upcurrents related to topographical features. I think the latter case illustrates a typical developing nocturnal inversion situation, in which no assistance whatsoever can be obtained from air currents. Rooks and Jackdaws hedge-hop against strong winds, but fly high and steadily to their roosts in still clear weather.

When I studied the large wintering population of Crag Martins in Gibraltar,[45] I found that their flight behaviour on return to the roost varied with wind direction and speed. The return was invariably in a southward direction over the coast to the west of the Rock. On all occasions when winds were light, the birds flew at up to 300 m above sea level. In strong westerlies, flight was direct but low, often at roof top height close in to the Rock. Strong easterlies resulted in a completely different behaviour. These winds create

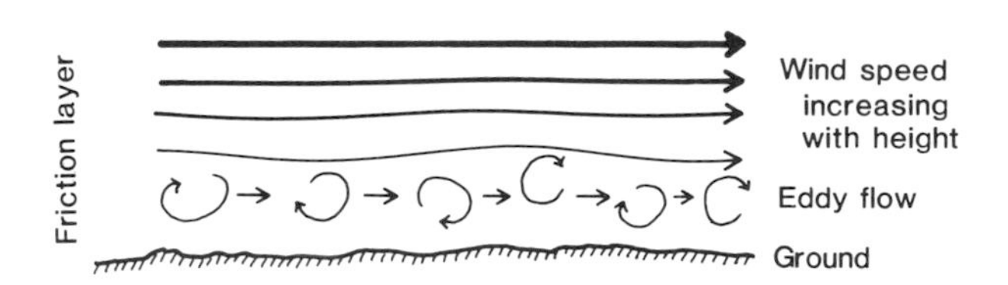

Fig. 11 Turbulent eddies formed in strong wind.

considerable turbulence to the lee of the 426 m-high Rock and the eddies are marked and powerful enough to give problems to large aircraft. At such times, the Crag Martins flew further from the Rock at heights up to 150 m, and there was a great deal of circling in the eddies.

Following contours along rising ground in a crosswind will ensure a supply of lift to minimise flapping. Cormorants and gulls will glide along a shore for long distances in the zone of rising air above unbroken sand dunes. If gulls fly out of this zone for any reason, the loss of lift forces an immediate resumption of flapping flight until a return to rising air. Long breakers on the sea shore give a similar zone of lift. Gulls can coast effortlessly, gliding rapidly with one wingtip almost touching the smooth water just below the wave crest, and finally using their momentum to climb rapidly as the crest breaks.

Lastly, a word about the difficulty of flight in very strong surface winds. Most birds probably go to ground to avoid the risk of collision with obstacles, but even so forward flight into wind seems possible. Col. R. Meinertzhagen quoted such instances,[104] including Eiders flying with a ground speed of 7–9 m/s against a wind of up to 40 m/s measured at 6 m above the ground.

FLIGHT ATTITUDE

Birds vary their flight attitude – the set of wings and tail – to suit the strength and gustiness of the airflow in which they move. In thermal soaring, the wings and tail are spread to ensure that the maximum surface area possible is presented to the rising air current. In orographic flow – where there is a greater horizontal component to the wind – wings are often bent, and feet lowered, with flight mostly at a lower level than in thermals. If upcurrents are weak, the body and wings are very stable, and the tips of the primaries can be observed moving incessantly as the current changes. The emarginated primaries of large soaring birds, with deep wing slots, are an extreme adaptation to static soaring flight. As each primary is spread, it bends under varying aerodynamic loads. The feathers lie one above the other – acting as individual aerofoils in the manner of a biplane or triplane. The shape of the wing slots is important, a U-shape with marked emargination being more efficient than a V-shape since it reduces turbulence and increases lift.[20] The stability and steering of a soaring bird is determined largely by the tail, which is in constant motion. The size and weight of a large bird gives it sufficient momentum to carry it through small erratic air currents without loss of stability or control. African vultures, if carried too high by strong thermals when hunting, lower their feet to increase drag and rate of sink.[135]

Many birds appear motionless in orographic soaring. Perfect poise and control is apparent in the aerobatics shown by some raptors and corvids. The Golden Eagle is arguably the most impressive flier of all, and many spectacular manoeuvres have been described – seemingly executed for no other reason than 'enjoyment' of flight. I have watched a bird rise in an orographic upcurrent then fold its wings to dive to a much lower level, whereupon, at considerable speed, it opened its wings suddenly to soar immediately upwards again. Seton Gordon[60] watched a Golden Eagle flexing its wings so that the tips were curved inwards to the extent that they almost touched. In this position, it remained stationary in a fresh breeze. Stiffening the wings to the usual soaring attitude enabled the bird to move forward into the wind without any wing thrust.

As mentioned earlier, height gained in rising air currents with little expenditure of energy can be used for distant travel. It enables a bird to glide for a long distance towards, perhaps, a distant hunting ground. The Golden Eagle may travel up to 22 km in order to hunt and its flight attitude varies according to the wind direction. The speed of gliding is controlled by the degree of wing flexing, and long effortless glides are a combination of skill and momentum. Against the wind, a bird slants downwards from its initial elevation with wings close to the body. This enables its momentum to carry it forward at an often impressive speed. In flight across the wind, occasional wing thrusts must be made. Flight with the wind requires a shallower gliding angle with straighter wings held further from the body. There are records of fast flight through hill fog with visibilities of 20 m. In these cases the birds doubtless detect air currents accurately enough to avoid collision with the ground.[60]

In the turbulent flow beside sea cliffs, the impressive exhibitions of flight control by the Fulmar are well known. The birds stall and turn with precision in the strong and rapidly changing upcurrents, often hanging motionless with angled wings, and steering with lowered feet. Nevertheless, J. B. Nelson[114] recorded an adult Gannet being blown head over heels over the cliffs of Ailsa Craig in severe turbulence, and I have seen a Fulmar tipped into a wave crest by a sudden gust.

When hanging in a strong wind, a bird actually glides in shallow descent at the same speed as the opposing airflow, so maintaining a stationary position. The head is held steady and level while the body and wings absorb the buffeting of the air currents. Lift provided by the airflow compensates for the gentle downward attitude, and is increased by the fanning of the tail. Some species such as the Buzzard, Peregrine and Red Kite usually hover over windward slopes where lift is more prominent. In a gusty wind, the stability of the Kestrel is controlled by rapid tail movements. The fast but shallow beating of the wings, and the angle of the body (between a horizontal plane and an angle of 45°) are a response to the strength of the airflow. Other raptors hover or 'hang' on the wind, and terns and the Kingfisher do so when searching for fish. Even Herring Gulls have been recorded hovering in updraughts to watch for and prey upon young rabbits. Harriers, with their low wing loading, are able to fly slowly at low level with a high degree of manoeuvrability, often at walking pace in a strong wind. More flapping is needed in light winds.

DYNAMIC SOARING

Another type of soaring, generally used over the oceans, exploits adjacent air currents of different velocities, and is known as dynamic soaring. Horizontal air currents possess considerable shear just above the earth's surface, due to friction, and this shear is prominent over rough seas in very strong winds. Dynamic soaring is practised by two families of seabirds, the albatrosses and shearwaters, and occasionally the Gannet in high winds. Flight by some of these species is among the most dramatic to be found.

Albatrosses have long narrow wings and are ideally suited to the very strong winds which blow with great constancy over the mid and high latitudes of the southern oceans. Except during the breeding season (and even then they generally use small islands or promontories that are almost surrounded by sea), their lives are wholly oceanic. The mid latitudes come under the influence of the mobile depressions (see Chapter 1) which create strong airflows, and nowhere is this more marked than between latitudes 40°S and 50°S. In this region there is little land to bar the passage of depressions, and it is not surprising that it is called the roaring forties. Calms are relatively rare, and it is significant that these soaring birds find flight difficult in calm weather. There is fossil evidence that albatrosses once inhabited the North Atlantic, and they may have died out through changing meteorological conditions.[25] Certainly, few cross the capricious winds of the doldrums. Occasional birds, chiefly Black-browed and Yellow-nosed Albatrosses, reach the North Atlantic, perhaps in the southerlies on the western flank of the high pressure zone.

When on land to breed, albatrosses nest on the windward sides of islands or promontories. They need only stretch their wings and legs to lift into the air on the strong updraughts over the rising slopes. In all but extremely light winds, albatross flight is closely linked to the marked wind shear immediately above the waves. As long as their airspeed is high, they are able to maintain height. During the prolonged gales in the expanse of the southern oceans, the significant wave height (the term used for the average of the highest one-third of waves) is around 8 m from crest to trough. Since ocean wave systems change only slowly with time and distance, eddy flow among the waves is marked, and considerably reduces the wind velocity in the troughs. A high airspeed is achieved by climbing rapidly from calmer air in the troughs into stronger winds aloft, assisted by upward components of the airflow deflected by the waves. This latter lift may often be a greater source of energy than that due to the wind shear itself. On reaching heights of up to 20 m, the birds turn downwind and glide at high speed, finally banking along wave troughs to repeat the process. The wings are flexed and twisted at the carpal joint to take full advantage of the air currents. In full gales the wings may be raked back steeply. With windspeeds below 5 m/s, flapping becomes more frequent, while in calms they prefer to settle on the sea, although slope-soaring over moving waves has been observed.[210] They have great difficulty in taking off from the sea or level ground in calm weather, often executing long but unsuccessful take-off runs. Some albatrosses, and certain other southern seabirds, regularly circumnavigate the globe on the southern westerlies in their pre-breeding years. They have also been recorded circling in hundreds up to an altitude of 1,500 m in sunny early summer weather, suggesting that they are able to take advantage of rising currents over their breeding grounds, either in thermals or, more likely, lee waves.

The smaller oceanic soaring birds of the shearwater family are equally at home in the windswept oceans, although the larger species are apparently more tolerant of high winds than the small Manx Shearwater. Their mode of flight is often similar to that of the albatross, except that they tower into the wind to a lesser extent, and show a more direct flight. In light winds they can travel satisfactorily by flapping and gliding. The Fulmar depends more on upcurrents deflected by the waves rather than wind shear i.e. it is more of a static soarer than a dynamic soarer.

SUMMARY

In this chapter, I have described the types of airflow that birds make use of. They can be grouped into categories, each used in a different manner. Convection is generated naturally and artificially over land and sea, but convective upcurrents, thermals, are used chiefly in static soaring by species inhabiting continental climates where solar heating is at its maximum and most prolonged. The lift produced by orographic flow over hills and mountains enables many soaring birds to live independently of thermals, and the turbulence and wind shear over ocean waves assists dynamic soaring by certain seabirds. I have also shown how birds have adapted to the use of these airflows, and how their flight attitudes vary.

I have not investigated – but will do so in succeeding chapters – the motives behind the use of the various airflows. The birds save energy, of course, and in addition some use upcurrents as travel aids, either between feeding grounds and roost or nest site, or for migration. Birds of prey, and certain other species, employ upcurrents for hunting or display, and I will describe later how convection is of immense importance to species which feed *in* the air rather than *from* it.

Finally, all who watch birds can enumerate many instances in which upcurrents are used for no apparent purpose and, on some occasions at least, flight in rising air currents seems to be undertaken for no other reason than 'enjoyment'.

CHAPTER THREE

Feeding

Except during the breeding season almost the whole of a bird's waking hours can be directed, if necessary, towards the search for food. By winter, any migratory journeys have been undertaken and (with no eggs to incubate or young to care for) individuals are able to forage over a much wider area than in summer, although some move little. In summer, however, adult birds not only need to feed themselves but also find sufficient food for their young. The foremost distinction between winter and summer feeding lies in the type of food available and its distribution, and the abundant populations of insects present only in summer in mid and high latitudes form the food resource of a very large number of species.

The amount of foraging or hunting that diurnal species can accomplish is controlled by daytime light intensity, which may be modified by meteorological factors such as thick cloud. The importance of daylength in winter increases with latitude, since at 66°N, for example, midwinter daylight (including twilight) is only 62% of that at 45°N. At this time it follows that any reduction in the ability of an individual to find sufficient food may prove fatal if the interruption is prolonged. It is the length of a period of adverse weather, rather than the intensity of the contributory element, that is significant. Provided that an adequate supply of food is readily available, and that foraging can continue, the survival of an individual and its offspring (in a meteorological context) is more or less assured.

It is in winter that the weather has the greatest influence on food-gathering. The effects of low temperature, snow cover, frozen water and ground and the icing of vegetation and other food sources are compounded if prolonged, particularly since low temperature increases a bird's maintenance needs, i.e. more fuel is required to maintain its metabolic rate.

EFFECT OF WEATHER ON FEEDING HABITAT

Temperature

A fall of temperature below 0°C results in a frost, but visible evidence will only appear if standing water or moist surfaces freeze, or if the air is moist enough to deposit white ice crystals (hoar frost) on the ground and vegetation. Precipitation falling at low temperatures is generally of snow, the formation of which was described in Chapter 1. The consequence of snowfall is a covering of the ground and low vegetation. Although level undrifted snow may cover all except bushes and trees, falling snow is frequently accompanied by wind which drifts it unevenly against obstructions, often leaving open ground on their lee sides. Therefore a considerable amount of snow must fall before the ground will be completely covered. The size, density and water content of snowflakes vary, depending on the temperatures and vertical currents within the snow-bearing cloud, but it is a general rule that 1 cm of rain is very approximately equivalent to 10 cm of undrifted snow.

Once a snow cover has become established, the temperatures of the air above the snow and of the soil beneath are considerably modified from those associated with bare ground. Since snow is a poor conductor of heat, it acts as a blanket to the soil, with soil temperatures higher under snow than on open ground. During the daytime, snow absorbs little of the incident solar radiation. It reflects around 85% compared with about 12% for moist bare soil, although enough is often absorbed to raise the snow surface temperature to near 0°C, unless of course the air temperature is well below this figure. The radiative effect of the snow surface is high, and under clear nocturnal skies the surface temperature drops to a markedly lower value than that of a snow-free area. A steep temperature gradient therefore exists within the snow at night, with the temperature increasing downwards. A hard crust is often formed on older snow. Bare soil thus freezes more readily than snow-covered (or even grass-covered) soil, although the temperature at which soil water freezes varies with the water content and soil type. Provided that the soil is unfrozen prior to deep snowfall, it will normally remain in that state beneath the covering so that the timing of frost in relation to snowfall is crucial to the state of the ground.

Snow melt is most quickly attained by the movement of mild air over the surface, or by the impact of rain, rather than by direct solar radiation, but the conduction of heat from warmed bare ground adjacent to the snow often plays a significant role. Additionally, living plants absorb solar radiation, whereby they melt shallow snow around them. The temperature of living buds may be as much as 10°C higher than the air temperature, and fruits and seeds can undergo a considerable degree of supercooling before they freeze.[54]

Deep snow cover prevents birds foraging successfully on the ground, except where clear areas exist in the vicinity of woodland and tall vegetation. Woodland functions as cover to protect its inhabitants from extremes of weather. Trees lessen the amount of snowfall reaching the ground, and ameliorate the effects of nocturnal radiation loss, so that the frequency of frost-bound and snow-bound soils is lower. Comparisons of winter temperatures have revealed that at night they are higher within a forest than in adjacent open country – more so in Norway Spruce than in Scots Pine, but only marginally higher in Beech woods. This reflects the density of the leaf canopy and its prevention of radiation loss from beneath. The age and species of tree are important, and open leafless Birch woodland is almost as extreme as open country. Within evergreen forest the diurnal range of winter temperature is lower than in open country by about 1–2°C. Furthermore, the amount and distribution of snow reaching a forest floor is determined partly by the type of snow. Wet snow is intercepted by the crowns of the trees, falling in large masses when sufficiently heavy, whereas powdery snow reaches the ground more easily and uniformly.[54]

A relatively uncommon winter hazard in Britain is glazed ice. Rain or drizzle falling on to surfaces whose temperature is below 0°C will freeze on impact to produce a complete coating of ice (Plate 3). In severe cases twigs and branches break under the weight of the ice. A covering of glazed ice on vegetation makes food inaccessible both to arboreal and ground feeders and, if persistent, mortality due to starvation may be high. It occurs for short periods at some time during most British winters, but is commoner in countries that experience more frequent subzero temperatures in conjunction with incursions of warm air.[96] A more common but less severe form of ice on vegetation is the white opaque deposit of rime. The sides of twigs and stalks exposed to the airflow gather an icy deposit in freezing fog, owing to the impaction of supercooled water droplets. In persistent freezing fog, accumulations of several centimetres may temporarily prevent feeding by birds and other animals.

The formation of ice on fresh water bodies is controlled by several factors of which the most important are exposure, wind speed, water depth, area, and force of inflow streams. Fresh water is unique among liquids in that its maximum density does not bear a linear relationship with temperature, but occurs at 4°C. Above this figure, the cooling of the surface layers induces a convection process, in which the colder water is continually being replaced with warmer, less dense water from beneath. The convective overturning ceases when the entire body of water reaches the temperature of maximum density. Further reduction in temperature to 0°C can then cause spontaneous freezing at the surface provided that there is no frictional mixing by wind. The cooling spreads slowly downwards, so that the bottom water freezes last. Deep water therefore takes a considerable time to freeze over, and unless frost is severe and prolonged, may not freeze throughout its depth. Ice forms firstly where the water is shallow and least disturbed, and where small solid objects project, although swimming birds may initially disturb the surface sufficiently to retard freezing. On fast-flowing rivers and streams, ice forms less readily, except that in prolonged frost some forms on the bed – adhering to rocks – and occasionally on the surface in small patches, where it eventually extends from bank to bank. In marshy areas, the invariably shallow water and wet ground freezes rapidly and completely. During the thaw, it is the ice on shallow water that thaws most rapidly.

Saline waters behave rather differently. The salinity of the sea depresses its freezing point, which at normal ocean salinity is −1.9°C. The maximum density of this water is at −3.5°C, and so convection begun during cooling at the surface will continue to much lower temperatures than in fresh water. Thus, ice forms less readily on the sea, but more rapidly at fresh water outflows. At all times, the density of the fresh water is lower than that of the sea, resulting in a layer of fresh water spreading out over the more saline water. The actions of wind, waves and tidal currents, however, serve to delay ice formation even further by mixing the strata of different salinities. Seas which are of low salinity – mainly enclosed waters at higher latitudes such as the Baltic – freeze readily. The Baltic Sea is of a salinity at which the maximum density is around 2°C; since this water freezes at just below 0°C, its character is more like fresh water, and in severe winters much of it freezes. Similarly, the low salinity of estuaries results in a more immediate freezing of the mud at low tide. I have walked over the mudflats of a small estuary which had been completely covered with a thick crust of ice – formed when overnight frost followed a covering of mixed river and sea water (see Plate 5b).

Wind

Although strong winds occur throughout the year, particularly in maritime climates, the greater frequency and intensity of polar front depressions in winter make them more noteworthy at that season. However, occasional deep depressions do affect higher latitudes in summer; and in regions subject to tropical storms, late summer is a period when these may impinge upon land areas.

Wind presents few problems for land birds during foraging unless it reaches great strength. There is little on record concerning the effects that very strong winds have on small passerines; they probably impose a limit on the height at which birds can forage successfully, since the decrease in wind speed can be very marked in the boundary layer close to the earth's surface. The habitat naturally determines the degree of shelter available, and those regions with maximum exposure – islands, coasts and high ground – pose special problems to the relatively few species that reside in them.

Within a forest, the wind speed decreases sharply from the canopy top, which acts as a friction layer. This wind shear is less steep on the windward edge, and also in winter among bare deciduous trees where wind can penetrate deeply. Thickets and bushes reduce the airflow to near zero, and the wind profile within woodland is also dependent upon the age and density of the trees. In clearings, large scale eddies are formed in strong winds, so that shelter may be reduced on the surrounding edges.[54]

Gales are usually worst on windward coasts where the reduction by friction over the sea surface is considerably less than inland. If strong enough – generally above 25 m/s – winds can cause widespread destruction of trees and possibly damage large areas of forest. The propensity of trees to windthrow depends upon height, species and site. Whole areas of trees can be flattened in areas lacking topographical shelter, but in irregular terrain destruction is confined to those trees lying in the path of winds strengthened by valley funnelling. Most healthy trees are firm against wind, but a great deal depends on the underlying soil and depth of rooting. Trees above 12–15 m in height are more liable to be felled than those below 6 m; thus birds of mature woodland are more likely to suffer loss of habitat than those inhabiting small plantations or shrubs (Plate 7).

PREY BEHAVIOUR

The effect of weather on prey is one of the more important indirect influences on avian feeding behaviour. Small mammals become inactive during snow and frost, and invertebrates which inhabit the surface of the ground or live in the topsoil also show a seasonal pattern of abundance. In cold or dry weather some withdraw to deeper soil where it is warmer or moister. Others remain motionless in the topsoil, even when it is still unfrozen, so that they are difficult to find. While soil temperature variation depends partly on the thermal conductivity of the soil, it is a general rule that the diurnal variation decreases with depth. In cold weather the temperature itself increases downwards, while the reverse is true in hot weather. Burrowing invertebrates can thus only be caught in abundance when extreme conditions ameliorate. Heavy rain, for example, brings earthworms to the surface, particularly after a dry spell. Both the number and weight of invertebrates diminish in winter and it has been found that stocks of insects in pine foliage late in this season are closely correlated with mean temperature.[55]

The activity of intertidal invertebrates also declines with temperature. On the shores of northwest Europe sea temperatures remain relatively high throughout winter; the lowest mean temperatures are found on the eastern shores of the North Sea. In Britain the minima are reached in February at river mouths and sheltered shallow bays. Even these, however, are no less than 4°C in the Wash and northeastern Irish Sea. Thus intertidal organisms – warmed as they are by the regular wash of the sea – are relatively more active in cold weather than their congeners inland. Nevertheless, during cold spells the upper shore may freeze except when covered by the highest tides. In exceptional winters (which will be discussed in Chapter 11), both sea and shore freeze, causing considerable mortality to prey and predator, particularly the invertebrates near the northern edge of their range.[27] The more northerly distributed organisms are remarkably resistant to short periods of subzero temperatures. The freezing point of the body fluid can be lowered by increasing its salt concentration. Many are also able to absorb heat readily, so that it is only on the rare occasions when sea temperatures fall below 0°C with little sunshine that widespread mortality occurs. Nevertheless, as the mud temperature falls burrowing animals dig deeper. Since organisms on rocky shores are unable to burrow, they must endure more rigorous conditions than those on sand or mud, though they can shelter under seaweed and stones. Mortality is normally greater on wide mudflats (where exposure to freezing lasts for longer periods) than on shelving shores.

Fresh water life becomes inaccessible in frozen waters, with specialised fish eaters migrating from regions where freezing is of regular occurrence. In certain summer weather also, fish become more difficult to catch, with downward movement taking place in both low and high temperatures, while choppy water, turbidity and heavy rain reduce visibility in the water.[131] The significance of these factors varies with the fish species. Marine fish behave in a similar manner, and are discussed in Chapter 12.

In summer there is generally little impediment to the acquisition of prey, yet there are meteorological factors that do affect the immense insect population. They include rainfall, temperature and air currents. The variation in vertical air currents apply more to the higher flying insects whose activity is crucial to aerial feeders and which will be discussed in the next chapter. However, for montane breeding birds, such upcurrents can be of importance. For example, snowfields close to breeding sites in the Scottish highlands are beneficial, even crucial, to Snow Buntings, and assist other small upland passerines in the search for food.[115] Thermals and anabatic winds from adjacent valleys carry myriads of insects aloft and strand them on snow surfaces, often in great abundance. They are then extremely easy to locate against the snow. A snowfield also provides a continual supply of moisture along its periphery where many hill breeding insects, mainly dipterous flies and craneflies, find moist ground on which to lay their eggs; this concentrates emerging insects. Meadow Pipits apparently catch craneflies more easily when the insects are less active and crawl along the ground, so that even cool weather may not be too damaging. After breeding, Wheatears and Meadow Pipits move to higher altitudes to feed on the insects stranded on summit snowbeds. Thus patches of snow vary in importance according to summer rainfall. In dry summers, areas with no snow hold fewer insects, limiting both avian populations and their breeding success. Though summer snowstorms can be hazardous in colder climates, snow-free patches remain near boulders and rocks, and the soil beneath the rocks remains unfrozen. Insects temporarily crowd into such spots, to be exploited by sheltering birds.[116]

The emergence of insect populations is chiefly correlated with warmth, often in combination with moisture. In cold, wet and windy weather insects become more difficult to locate due to inactivity and a reduction in numbers. This leads to delays in avian reproduction in cold springs and breeding failures in summer. Conversely, these conditions minimise soil drying and lead to a continuous and more readily available supply of some soil invertebrates, and if persistent, prolong the breeding season for species exploiting such food. The organisms that live in the cool humid environment beneath vegetation are at risk when moisture is depleted and plants become desiccated in hot dry weather. In wetlands, evaporation in such conditions may increase salinity, also leading to fewer invertebrates. In woodland, however, the leaf litter remains moist with the highest humidities near the forest floor, so that in droughts invertebrates are easier to find than in more exposed sites. Flying insects in woodland concentrate in the upper crowns of trees where solar heating is at a maximum. Lower down, the rise in temperature and therefore the maximum flying insect concentration depend on the density of the foliage.[54]

FEEDING BEHAVIOUR

Woodland

Grubb[65] maintained that for woodland birds, the height and rate of foraging, the tree species frequented, the ground layer composition and the horizontal distribution of individual birds are all weather-dependent. Passerine species which remain in woodland during winter consist primarily of insectivores which feed amongst leaf litter, on bark, or in conifer foliage, and seedeaters which take fruits or seeds either from the trees or the ground. The numbers of the latter birds depend to a large extent on the success of the previous autumn's fruiting.

Familiar sights in otherwise silent winter woodlands are the mobile flocks of tits, which frequently also include Treecreepers, Goldcrests and Nuthatches. Some of these feed continuously on insect eggs and pupae hidden in bark. Other woodland birds turn increasingly to invertebrates as the stocks of fruits and seeds diminish, so that a prolonged spell of freezing weather when supplies of these are at their lowest (in late winter or spring) can have serious results. Even a return to mild weather in late winter is not always beneficial, since germination of seeds may render them inedible, though the increasing abundance of insects may counter this loss. In prolonged snow and frost, ground feeders reach starvation level before arboreal feeders. Some degree of excavation is possible, of course. Hawfinches take fallen fruit stones in winter and have been recorded excavating for cherry stones through 2½ cm of snow. Larger birds such as Pheasants scratch through snow and frozen litter in the search for food, and Robins have developed a commensal habit in which they follow birds and mammals (including man) to take advantage of soil disturbance.

Changes of feeding site result when food stocks become inaccessible. The ultimate is reached in the mass movements which will be discussed in Chapter 11, when whole populations make often considerable journeys to avoid starvation. Small-scale movements are continually in progress. In low temperatures and strong winds (when the movement of crowns of trees makes food gathering difficult) tree-feeding woodpeckers and tits descend, and feed more on the ground.[65] Robins do likewise, showing sexual differences in foraging behaviour. Changes in behaviour with falling temperature are most marked in the male, bringing it into greater competition with the female. M. East[41] thought that this increase in competition may have played some part in the evolution of separate winter territories. Robins also tend to move to woodland edges as food stocks diminish and, in Britain especially, often move out entirely into close proximity to human settlements.[86]

Individual birds defending territory in any habitat are able to ensure for themselves a more regular supply of food from a familiar area, though it is only during food shortages that the value of the territory becomes important. The Robin's aggressiveness on its winter territory is well-known. However, in very cold weather when food is scarcest, there is a tendency in this and other birds towards a breakdown of the territorial instinct, as other, perhaps non-territorial, individuals of the same species forage further and encroach upon neighbouring territories. Territorialism is suppressed by the need to forage more earnestly owing to increased food requirements.

In many areas, particularly those which regularly experience total snow cover, birds store food within their territories. In Sweden Jays and Nutcrackers can locate stores beneath a considerable depth of snow – 45 cm for the latter species. In Britain the Coal, Willow and Marsh Tits store food more regularly than other tits, especially in spring and autumn, though such food may constitute only a very small part of their winter diet. Stores are located on the ground in Britain, but in Scandinavia, where storage is commoner, more are hidden in bark where location is easier during periods of deep snow.[137]

In summer, few weather-related feeding problems arise in woodland. Those that do arise are almost entirely due to the inactivity of insect prey in cool wet weather. In the woodlands of central Europe, Collared Flycatchers descend to the ground only when food becomes scarce. Many young die in prolonged rain, even when a rich food supply still exists, since few adults appear to recognise motionless insects as prey.[94] Buxton[21] found that Redstarts gathered aphids from the ground where they had been washed from the trees in heavy rain.

Open country

Shelter in habitats outwith woodland is greatly reduced. This section and the next discusses the feeding behaviour of birds which cannot rely on the unique microclimate that woodland affords. In more exposed habitats the majority of ground feeders manage to reach food as long as snow cover is shallow. In many cases birds can benefit considerably from man-made food supplies, whether winter crops, stubble, sewage effluent or food waste of various forms.

The value of stubble fields as a food source is directly affected by the amount of rain falling during the summer and autumn. High rainfall on standing crops delays harvesting and causes more grain to fall to the ground; grain also falls from wind-beaten crops and, with ploughing delayed by rain-soaked ground, a large amount of grain is available over a long period to granivorous species such as the Woodpigeon and Greylag Goose. In a dry autumn the crop is harvested quickly and the stubble ploughed over as soon as possible, so that grain stocks may be finished by late October.[111]

Between 1970 and 1984, the Garden Bird feeding survey organised by the British Trust for Ornithology listed 125 species taking food at over 188 garden feeding stations throughout Britain. The number and variety of birds depend largely on the severity of the weather and the amount of food available naturally. The number of Great and Coal Tits, Chaffinches, Siskins and Jays visiting feeding stations varies not so much with the weather, but more with the abundance of the previous autumn's fruit and seed crop. Coal Tits show a definite biannual feeding activity pattern unrelated to cold weather, at variance with birds in pine forests which feed chiefly on insects.[55] It appears that, for seed-eating tits, winter survival in Britain is not closely correlated with the weather unless unusually severe. Tits tend to use well-established natural feeding sites as long as there is enough food, but in heavy snowfall desert them and move to gardens. In Oxfordshire, tits returning to woodland after cold spells and in spring had plumage soiled through feeding in urban areas.[137]

In general, the more rigorous winter weather in uplands forces most resident birds to lower levels. The degree of movement is related to the severity of the weather and its penetration from higher altitudes. Red Grouse, however, do not often suffer acute food shortages from winter snows, but make local movements to clear areas. Similarly, Ptarmigan congregate where snow has drifted off vegetation, or will uncover plants buried under a depth of snow.[179] In Alaska, Sharp-tailed and Ruffed Grouse starved when glazed ice covered the snow with a hard crust, preventing them from reaching the vegetation beneath.[103] Other grouse exchange open areas for woodland in deep snow cover. Altitudinal movement in montane species may be reversed when hills and valleys are both totally snowbound, since snow invariably lies deeper in low lying parts where less wind scouring takes place.

Summer drought is a greater problem in exposed habitats than in woodland, since the ground becomes dry and sunbaked. Species exploiting soil invertebrates find difficulty in feeding young. In such weather fewer young Song Thrushes starve than do Blackbirds because they are fed on snails rather than earthworms. Both the Song Thrush and Blackbird are primarily woodland species but the latter has expanded more into gardens. Although this has allowed it to exploit a greater variety and amount of food, leading to higher populations in the latter habitat, it is the woodland Blackbird which suffers less in a drought, surviving on caterpillars and leaf litter invertebrates rather than earthworms and insects.[162]

Birds of prey take alternative vertebrate prey when their preferred food becomes scarce. However, the vulnerability of small mammals increases in floods and when sudden temperature drops in previously thawed areas cover their tunnels and runways with ice. Atypical behaviour may occur in attempts to satisfy hunger, such as the daytime hunting by nocturnal owls. Glue and Nuttall[58] found that severe cold weather halved the number and nutritional value of prey items of an individual Barn Owl, which survived a week of frost and snow on a reduced mammal diet. Predators (including crows and the larger gulls) may actually benefit from the weakened state of their prey, but with the lower nutritional value the benefit is only of short duration.

Strong winds and continuous or heavy rain may delay or completely prohibit hunting. There are those birds which find a strong wind advantageous, such as the Hen Harrier and Short-eared Owl, although if very strong and turbulent, manoeuvres become laborious.[182] Some large raptors appear to dislike flying with wet plumage,[80] probably because efficiency is reduced to a level at which energy expenditure in flight becomes excessive. Rain also restricts vision and hearing and makes prey less active[182] and indeed, the acoustic properties of calm frosty weather raise the capture success of those owls which rely heavily on hearing.[207]

Buzzards are versatile hunters, modifying techniques according to habitat. Forest-dwellers hunt frequently from perches, seldom hovering; birds in open country do so only when the weather inhibits hunting. These and some other raptors hunt at low levels so that high level thermal and orographic soaring is less associated with feeding than with display or other activity.[20]

There is scant documented evidence to show that fog affects feeding. Certainly the hunting of at least the diurnal birds of prey must be inhibited, particularly in the maritime mountainous areas of northwest Europe which experience frequent low cloud. There are often prolonged periods in which dense hill fog envelops all higher ground, with visibilities less than 100 m. Difficulties must also arise in the dense fogs which occur over flat lowlands in the stable atmospheric conditions of winter anticyclones, particularly during the short winter daylight. Peregrines have been recorded taking mainly pedestrian species (e.g. Moorhen and game birds) in Essex during misty weather. Summer sea fogs in coastal areas can persist for days, perhaps also posing problems for coastal Peregrines.

Wetlands

With the freezing of fresh waters in northern Eurasia and North America, most wildfowl make lengthy migrations from their northern and eastern breeding grounds to avoid the prolonged winter. Only a few marine species and a small part of the Icelandic wildfowl population remain; the latter where thermal springs keep interior lakes free of ice. Trumpeter Swans do likewise in one area of the USA. In more temperate climates, the movement of wildfowl from fresh water to marine habitats becomes prevalent when there is extensive ice inland. Some species are more reluctant than others to forsake fresh water, resulting in enormous flocks on the few open deep waters. Even surface feeding ducks may then dive for food. Some diving duck feed below the surface of the ice as long as unfrozen areas provide access, and other aquatic birds, such as Dippers and Kingfishers, also do so. The latter has been recorded fishing through a very small hole in ice while ignoring nearby open water;[155] but in total ice cover, British birds either move to estuarine habitats, or die inland from starvation – the heaviest mortality taking place in January.

Dippers move from smaller ice-covered streams to larger waters or tidal estuaries and seashores. They will also dive from ice floes on occasions. Shooter[154] recorded that Dippers in Derbyshire only forsook their territories in severe frosts, particularly on streams which emerged from limestone, since these were 2°C warmer than those on gritstone at the same altitude. In southern England, some Moorhens defend what have been termed 'core' areas – winter territories coincident with, but smaller than, those held in the breeding season.[193] While one member of a pair defends the core area, the other may feed elsewhere, often with winter flocks. When waters are frozen, the areas become neutral and are no longer defended.

In strong winds little shelter is available on large water bodies; here difficulties in foraging are related to feeding habits. Surface feeding water birds are often found on smaller less exposed waters. Diving birds are less affected except in shallow waters where wind stirs up bottom sediment which reduces visibility. However, such disturbance of the seabed may expose new food resources, and lead to a redistribution of sea duck. In some situations there can be a dramatic reduction in the food supply. Persistent easterly gales in January, 1963, compounded by exceptionally low temperatures, caused heavy mortality to marine life on the east coast of Ireland. The food scarcity resulted in a marked decline in the wintering population of Common Scoters.

Inland, the inactivity and downward retreat of invertebrates in cold weather force wader and gull flocks to move to estuarine sand and mud. In Britain, the foremost inland feeders include the Oystercatcher, Curlew, Lapwing, Golden Plover, Snipe, Common, Herring and Black-headed Gulls and, in some localities, the Redshank. At the onset of frost or snow, the decline of the two plovers is not immediate, since the birds locate other suitable local feeding areas rather than waste energy by moving from the area during what might only be a temporary cold spell. Mass emigration takes place later if the hard weather becomes prolonged. Snipe, relatively solitary birds in normal feeding conditions, have occasionally been observed in very large flocks (up to 6500) where extensive freezing of inland floodings has concentrated them in favourable but localised feeding grounds.

Female Oystercatchers are more successful at feeding in low temperatures than males, since they are able to probe more deeply with their longer bills. A study of birds feeding in winter on coastal fields in Lancashire revealed that 97% were adult females, while the bills of the males present were significantly longer than those of the adult male population as a whole. Thus in respect of inland winter feeding, selection may favour birds able to reach the deeper burrowing prey.[28] The presence of grass fields is of great importance in midwinter to both Oystercatcher and Redshank (where the latter feed inland). In Scotland the former are unable to obtain sufficient food from estuarine sources, and will follow the plough if grass fields are frozen hard. At least a quarter of the population in one area was found dead of starvation in February 1969, when deep snow lay for 22 days.[71]

Estuarine invertebrates may become so inactive in cold weather that the energy used by waders in foraging might easily be more than that gained from the food captured. As the substrate temperature falls, waders take different food items.[61] The long-billed species are probably less affected than the short-billed, whose food items are the first to succumb to low temperatures. Each bird covers less ground in order to reduce energy expenditure. Those feeding more on the upper shore, such as Ringed and Grey Plovers, suffer proportionately greater starvation. Frozen mud also generates problems for the Shelduck, whose principal food item in Britain is the gastropod *Hydrobia ulvae*. Thompson[172] observed them scraping barnacles from wooden stakes in the Clyde estuary when the mud was frozen, and concluded that (like many other species) Shelducks may be opportunist feeders in severe weather. In such conditions, many estuarine waders move to rocky shores where shelter and the continued availability of non-burrowing organisms becomes very important.

Wind and rain also reduce the activity of some burrowing invertebrates, and prolonged onshore gales cover feeding grounds with unusually high tides. Many species of wader vacate exposed coasts in winter when rough seas are frequent. Those that do remain on rocky shores, such as the Turnstone and Purple Sandpiper, may not find sufficient food. A study of the winter feeding behaviour of the Purple Sandpiper[49] revealed that on stormy days, the extent to which the birds needed to evade waves increased to almost five times that on calm days. Feeding time may be reduced by as much as a quarter on short winter days. When seas are rough, some flocks are driven to alternative sites where they take different food.

Whereas most waders prefer to feed visually, and disperse to minimise interference from adjacent individuals, the need to seek shelter from strong winds and blowing sand may concentrate them into small areas where they are compelled to feed more by touch than by sight, thus having to take less preferred food items. There is evidence that when wind speeds exceed about 12 m/s, a few species experience difficulty in feeding. In particular, Redshanks appear unable to tolerate buffeting as much as other waders.[36,62]

An excess or a deficit of rain may modify feeding behaviour. Heavy prolonged rain in active slow-moving depressions, or the sudden downpours from vigorous convective cloud, cause flooding, both by depositing large amounts of water in low-lying areas, and by the overflow of streams, rivers and lakes. A similar effect is achieved by the sudden thaw of snow in rapidly rising temperatures. The mass of water deposited in wet weather is quite remarkable. Over an area of one square kilometre, a wet day with a mere 10 mm of rain will deposit 10 million kilograms (equivalent to 2.2 million gallons) of water.

Inland feeding waders and gulls take advantage of the increased abundance of organisms in flooded areas and are often accompanied by surface feeding ducks. Fluctuations in water levels from winter to winter result in variations in the distribution of wintering wildfowl. In eastern England, Wigeon feeding on the Ouse Washes adapt to prevailing conditions. Like geese, Wigeon prefer to graze on dry land, but when shallow water covers the feeding area they upend to reach submerged vegetation. They seek higher adjacent ground if the water is too deep. In summer, too, damp feeding areas are essential. The Redshank, for example, leads its chicks to the nearest suitable wet area as soon as they leave the nest. In a dry summer this may result in a long and perhaps hazardous trek for the family.

AGGRESSION AND PHYSIOLOGY

Further behavioural changes related to feeding concern aggression and adaptations to a bird's physiological condition. Aggression is shown, chiefly, by gregarious species. In addition to guarding against predators, flocking is thought to serve as a means of communicating information within the flock regarding food sources. This has great value in severe weather which, if extreme, may even induce flocking in those species such as Blackbirds, which are normally solitary feeders. When prolonged snow or frost reduces the availability of food or restricts the size or number of feeding sites, flock size increases while the distance between the individuals within the flock is reduced. In several species this elicits aggression towards the same or other species. The quarrelling and fighting of Starlings is a well-known and classic example, in which the close proximity of a neighbour stimulates the aggression. In this species, aggression in a flock beyond an optimum size tends to reduce the amount of food taken by an individual. Not all birds exhibit aggression under such circumstances, however. There are others whose aggressiveness decreases with both low and high temperature, probably because of the need to spend more time searching for food or shelter.

Clearly the aggressiveness of a species is dependent upon several factors, of which perhaps density is of prime importance. Some degree of peck order in feeding flocks is of survival value to all members of the flock, although in large dense flocks relationships may not be stable. For dominant birds, food can be obtained with the minimum of fighting, and more submissive individuals waste little energy in fights which they would lose anyway.[87] In marked food shortages, the less able individuals will die of starvation if they remain with the flock, although in cold midwinter weather it is less advantageous for birds to retreat from food during combat, so that fights may increase.

Physiological changes are largely caused by fluctuations in environmental temperature, affecting weight and fat reserves of individual birds. The metabolic rate of a bird is high and in order to maintain its body temperature at an optimum level the rate increases with falling air temperature. The greater the difference between body and air temperature, the more rapid the heat loss. A larger intake of energy is therefore required, brought about by an improvement in an active bird's appetite. Alternatively energy reserves within the body are utilised. Diurnal weight changes vary according to a bird's feeding habits. The behaviour of many shorebirds is regulated by the tide, so that they experience a shorter break than those birds feeding diurnally, whose greatest loss of weight occurs at night.

While losses are moderated by adopting certain behavioural strategies (see Chapter 6), they nonetheless increase in cold weather. Survival is enhanced if the day's food intake not only replaces the loss, but also adds some reserves of fat. Fat deposition can be achieved as long as a plentiful supply of food is available, and in winter peak weights are generally reached prior to the seasonally leanest period. Should a food scarcity then follow, these reserves may mean the difference between survival and starvation[168] and will tide individuals over a short lean spell. If the scarcity is prolonged, then weight is lost rapidly as the reserves are utilised, and mortality ensues. In severe weather, weak and emaciated birds are an all too common sight. Once a bird has used up its body fat, other tissues begin to be metabolised and the bird reaches a point from which it cannot recover. Small birds in particular require considerable energy in low temperatures since their surface area is large in relation to their volume, and heat loss is proportionately greater than in large species. Many generally store only enough fat to survive overnight, although some can accumulate a modest reserve in winter. Very small passerines may survive for as little as 4–5 hours after gizzards are empty, but large birds are able to exist for much longer. Pheasants can endure 2 weeks without food, and Grey Partridges 4–5 days. During extreme cold, small birds alternate between intense feeding and resting, while others may feed for the whole time available.

Energy is used not only to maintain body temperature, but also for flight. American studies of Carolina Chickadees found that the species changed its feeding behaviour as temperatures fell below 0°C.[112] These small-billed birds normally eat small sunflower seeds, but move to large seeds in cold weather to increase the amount of food taken per unit effort. The time taken to eat an individual seed doubles, but fewer journeys to food plants result in a net conservation of energy. Repeated disturbance of any feeding birds in cold weather may have serious results, since the energy used in unnecessary flight may be critical.

Many studies of weight variation have found correlations between weight and temperature. The winter weights of most waders in Britain reach a peak in December. At that time mean weights are inversely correlated with the mean temperature. For example, weights of Dunlin are higher in eastern areas than in the milder south-west. This variation is assumed to be a form of insurance against possible food scarcity later in the winter and, indeed, peak mortality does occur in some wader species in late winter.[139] However, recent work in Britain has indicated that, for waders feeding in open habitats, fat deposition may be more of an insurance against gales than low temperature. There is evidence that wind chill and the curtailment of feeding by strong winds have marked effects upon the weights of some larger shorebirds during periods in which these factors are most severe.[36]

Bullfinches and Greenfinches also lay down extra fat in winter, both as insulation and a short-term food reserve; they do so by eating more in cold weather[121] (Fig. 12). However, the Bullfinch's accumulation of fat *during* cold weather contrasts with the strategy employed by many other species which gain weight before temperatures fall. In the extremely low temperatures of 1962/63 (see Chapter 11), very high Bullfinch weights were recorded in England, but individual values were variable, with some birds not reaching the appropriate survival weight. During food scarcity, a variable distribution

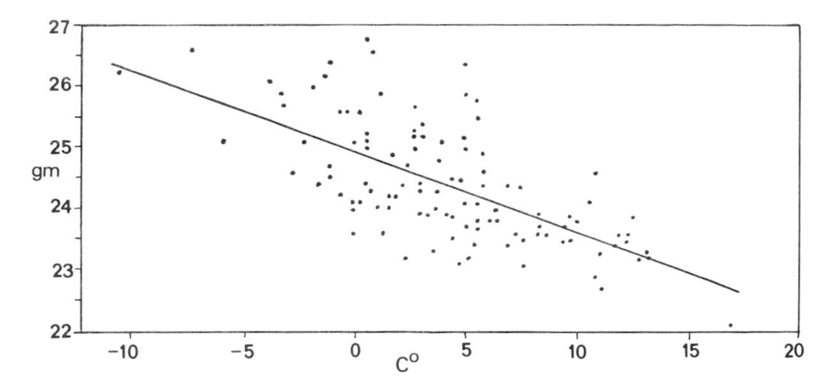

Fig. 12 Weights of Bullfinches in relation to air temperatures. Each point is the mean of at least ten weights obtained throughout a single day. The air temperature for each day was taken as the mean of the maximum and minimum (from Newton 1972)[121].

of weight commonly occurs among individuals of gregarious species. Those birds which are successful in disputes are able to maintain their weight, whereas the losers are normally the first to succumb to starvation. In a situation in which the food scarcity is abrupt and marked (as in a heavy fall of snow after mild weather) there may be a large number of birds without reserves, and a greater risk of mortality. For seed-eaters, however, this rarely happens, and below a temperature threshold of about 12°C Bullfinches respond very rapidly to marked temperature changes and adjust their weights within a day or two. There is also the possibility that the weight increase of some birds could be due to fluid retention (nutritional oedema) brought about by the cold weather. In very cold conditions, the scarcity of lightweight individuals suggests that when the fat reserves have gone, death soon follows.[120]

Other finches are able to add fat reserves, and threshold temperatures are lower in colder climates. The extreme is reached in central Alaska, where Arctic Redpolls do not gain weight until the air temperature falls below −12°C. This species winters from Alaska to northeast Greenland and remains further north, in lower temperatures, than any other songbird. To assist survival in night minimum temperatures as low as −67°C, the feathers are designed for extra insulation and the larger gullet enables storage of food prior to roosting.

In several species, the digestive system is modified in winter to aid an alteration in diet. In some populations of the Bearded Tit, notably the more continental ones, small stones are taken into the gizzard (in which the lining will have hardened) to assist the crushing of seeds. The reversion to the softer summer lining takes place in February or March, and may be hastened by a spell of mild weather. Should this be followed by further cold weather, the birds may not then be able to consume the winter food, and starvation ensues.[163] Similar changes occur in some Blue Tits. Some species of grouse also take grit in late summer and early autumn to assist in crushing the tougher vegetation consumed in winter. Covering of grit supplies by early autumn snows, or late spring snows (when further grit is needed to replace the worn autumn intake) causes digestive problems and starvation.

SUMMARY

In this chapter I have described the influence of various meteorological elements on feeding behaviour. Prolonged wintry weather inhibits both foraging ability and the availability of plant and animal food. The activity of prey diminishes, making location more difficult. Changes arise in diet and feeding areas, and the relative differences between various types of feeding habitat in winter and summer are discussed. Many species resort to food sources associated with man, especially in winter. Behavioural and physiological adaptations resulting from cold weather include aggression and adjustments to internal fat reserves.

Snow Bunting pecking insects from snow

CHAPTER FOUR

Aerial feeding

In Chapter Two, I described the formation of thermal upcurrents and how birds have adapted their behaviour to make use of them. I emphasised that, in mid and high latitudes, species that depend upon thermals alone for hunting and foraging are almost entirely migratory, and must necessarily spend the winter in lower latitudes where thermals are generally available throughout the year. Since species which feed on the wing are chiefly insectivorous, with the exception of some of the smaller raptors, their feeding success is frequently determined by mechanisms which concentrate insects into discrete zones; and because airborne insect activity is similarly confined to the warmer parts of the year, these insectivores are also migrants.

There are relatively few species in this category, but the effect of weather upon them is very marked. Nightjars, swifts and hirundines are the chief families involved, although if there is an abnormal abundance of airborne insects other species less adapted to aerial feeding may also take advantage of this food supply. Since their methods of feeding are much the same throughout the year, I will discuss behaviour both in summer and winter quarters.

THE FOOD

Studies of aphid distribution from ground level to a height of 600 m[76] have shown that insect density is greater at higher levels when the air is unstable. Since these insects have a very feeble flight – of 1 m/s or less – they are very easily displaced by even weak air currents. After a day of instability, there is a rapid descent of insect swarms as cooling begins and stability sets in. This is often clearly evident on a still evening when swarms suddenly appear at low levels. The aphids concerned are those species which seek distribution over a wide area, and thus tend to migrate in windy conditions. Many insects require warmth in order to fly, and are wholly dependent upon the air temperature since they cannot produce sufficient energy themselves. The threshold temperature at which they begin to fly differs among species, but once that threshold is exceeded, the proportion of insects in flight is independent of temperature until a high upper threshold is reached.[169] Nocturnal insects such as moths, which form a large part of the diet of nightjars, are large enough to be able to warm their bodies by muscular contraction, and so require a considerably lower threshold temperature than diurnal insects.

Upcurrents in warm unstable airmasses tend to concentrate aerial plankton in the form of flying insects and minute spiders. The one family to exploit these concentrations in vast numbers is the swift. It is notable that their abundance in many instances is such that echoes attributable to both birds and insects appear on radar screens. The most intensive activity of the Swift in Britain is normally in July when the young are in the nest. They are fed mainly on dipterous flies and aphids.[88] In the south of England, airborne insects begin to become abundant in April, and decline rapidly after September. Their density is initially dependent upon the state of emergence at ground level – whether hatching from vegetation or water – and finally dependent upon air currents. Due to the variability in emergence, their abundance is therefore not clearly correlated with meteorological factors. Though wind and temperature lapse rates are of importance in distribution, the duration and intensity of rain is also significant. Marked reductions in abundance take place in prolonged rain, particularly as temperatures frequently show significant falls in such conditions.

SEA BREEZES AND SQUALL LINES

In Europe, thermal upcurrents are frequent in summer, though perhaps less vigorous in the more maritime climates of the northwestern seaboards. There are also two other important sources of rising air. These are the sea breezes, and the gust fronts (or squall lines) associated with thunderstorms. The former are essentially coastal, while the latter are more continental in distribution, although not infrequently found in coastal areas.

Sea breezes are formed when a temperature contrast exists across a coastline with the higher temperature over the land, and a light general windflow. They are, in detail, rather complex daytime summer features

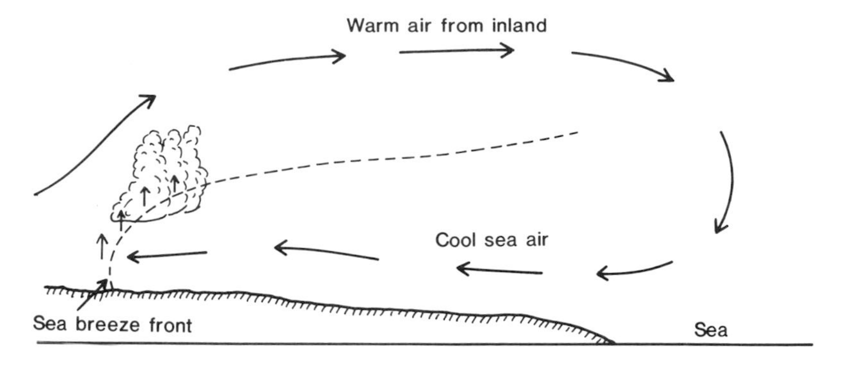

Fig. 13 Simplified cross-section of airflow in sea breeze circulation.

created when the land is heated more than the sea. A sea breeze results when the expansion of warmed air overland generates a high level outflow from the land and a compensating low level inflow of cooler air from the sea. The breeze is normally at its strongest by mid afternoon, provided that the overall meteorological situation remains unchanged. Its inland penetration depends upon the intensity of solar heating and the strength of the overlying airflow. In Britain, its maximum penetration is probably in the order of 80 to 150 km in southern areas, but it may reach up to 240 km in the tropics.

The cool sea air converges with the warm air inland, and acts as a wedge beneath it in a manner similar to that in a cold front. If there is sufficient moisture available, the lift which forms in the convergence zone is marked by a line of Cumulus cloud (Fig. 13). Should the general airmass be very unstable, and suitable for large scale convection, then Cumulonimbi may be formed with a possibility of thunderstorms. The maximum vertical depth of the cool air is, on average, about 500 m in Britain, but has been detected up to 1200 m and is perhaps deeper in the more unstable airmasses. There are cases in which a similar convergence zone is set up beyond the edge of a persistent layer of low cloud, which itself acts as a 'cool sea'. If the airmass is relatively stable and dry, the sea breeze front is often marked only by a change in horizontal visibility, with considerable haze in the sea air. Nevertheless, whatever the visible manifestation of the rising air, some species of birds must recognise it for what it is – a potential feeding area. The zone of lift gives upcurrents in the order of 1 to 3 m/s, rarely more except where enhanced by convective cloud, and it is in this region that airborne insects are concentrated.

While birds such as raptors and gulls will soar in these zones, the most abundant avian species associated with sea breeze fronts is the Swift. Many observations have been made by glider pilots who have found themselves sharing the rising air with these birds. In southern England, pilots have observed groups of Swifts feeding up to an altitude of 900 m.[157] Although occasional Swifts have been seen soaring under such circumstances, they are ill-adapted to this mode of flight, and more usually show a jerky, darting flight indicative of feeding. With a possible aphid density of two per cubic metre in the rising air of a convergence zone,[76] Swifts can easily collect sufficient food, and it has been calculated that they may catch an insect every five seconds during a ten hour feeding day.[158]

Swifts have been detected by radar in sea breeze fronts, moving along lines of convection in 'streets' of thermals, and in association with thunderstorm activity. As well as containing strong upcurrents, thunderstorms produce downdraughts of cold air which spread out at ground level. In some thunderstorm clouds, particularly those in which the wind changes little in direction or speed with respect to altitude, the downdraughts – often coinciding with falling precipitation – effectively counteract the upcurrents. Thus the thunderstorm begins to dissipate after the precipitation has begun. In other storms, a marked veer of wind with height ensures that updraughts and downdraughts can exist in separate portions of the cloud, and the storm can therefore maintain itself for long periods. The downdraughts from thunderstorm clouds are sometimes well-marked, especially ahead of a line of storms such as might occur along a cold front. There is often a zone of lift where the outflow of cold air acts as a front itself – known as a 'gust' front (Fig. 14). These vary in intensity and are typically up to 1 km deep. The front of cold air undercuts the warmer air ahead, and may trigger further convection downwind of the original storm – thus propagating shower activity forward with no necessity for new supplies of thermals from the ground.

W. G. Harper[67] investigated the origin of certain echoes – often intense – on weather radar screens, with simultaneous observations through a telescope fixed to the radar aerial. He discovered that the echoes, in the form of lines and arcs, were of parties of feeding Swifts that apparently were feeding upon flying insects concentrated in the zones of converging air. He found that the echoes in the form of lines were of Swifts in thermals associated with Cumulus cloud 'streets', in which the birds broke away from the upcurrents before reaching the cloud base. These occurred exclusively in fine warm summer weather. On the other hand, irregular arcs were associated with more vigorous convection. They were sometimes 25 to 40 km long, and lay parallel to, but several kilometres from, the edges of echoes from the shower activity of Cumulonimbus clouds. Some echoes were observed to ascend 300 to 500 m over a period of several minutes, and then descend. The arcs were clearly linked to the convection associated with gust fronts, and often preceded the appearance of echoes due to showers forming in that convection. The Swifts were exploiting insects carried aloft in these upcurrents, avoiding the prolonged and heavy precipitation normally associated with the cold downdraughts, which contain few airborne insects (Plate 10). This does not mean that Swifts do not fly in thundery showers. I have watched parties of Swifts feeding in heavy rain at the edge of a vast August thunderstorm, where there was obviously sufficient food in the warm calm air away from any

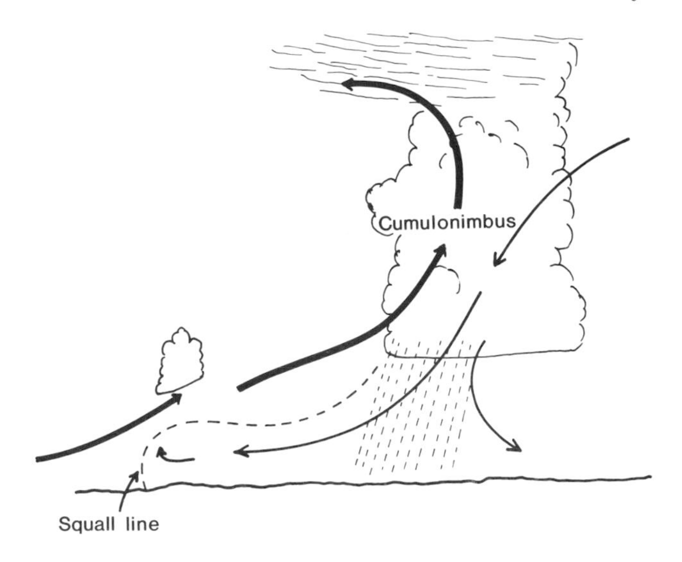

Fig. 14 Gust front (or Squall-line) associated with thunderstorm cloud (simplified). ➡ Warm Air; → Cold Air.

downdraught. It is of interest that in parts of Europe, the Swift is known as the rain- or thunder-swallow (cf Swedish 'tornsvala').

One particularly interesting example of Swifts feeding in convergence zones was observed on radar in July 1966.[148] At the time, an area of thundery rain over southern England had initiated a low level outflow of cool air northwards into East Anglia. Prolonged sunshine in this region had raised temperatures sufficiently for a sea breeze to form and move southwards from the Norfolk coast. A slack pressure distribution gave very light winds and allowed these fronts to dominate the low level airflow. In the early afternoon, two lines of radar echoes were seen to approach each other and finally cross. The nature of the echoes led the observers to interpret them as strings of Swifts feeding in the lift produced by each front of cool air. On meeting, the colder of the two airmasses – that due to the thunderstorms to the south – undercut the sea air. At this point, the extra lift was enough to form Cumulus cloud and eventually showers, which produced echoes still moving south with the upper level wind. The line echoes, however, merged into one line – the Swifts all having transferred their feeding activity to the still cloudless and dry northward moving front (Fig. 15).

Thunderstorms may also occur at levels above, say, 2 km, without any thermal activity. These are triggered by convergence in the middle troposphere, usually in conjunction with a cold front or trough of low pressure. Since there is no mechanism which can initially lift insects aloft, such storms are not suitable for foraging Swifts.

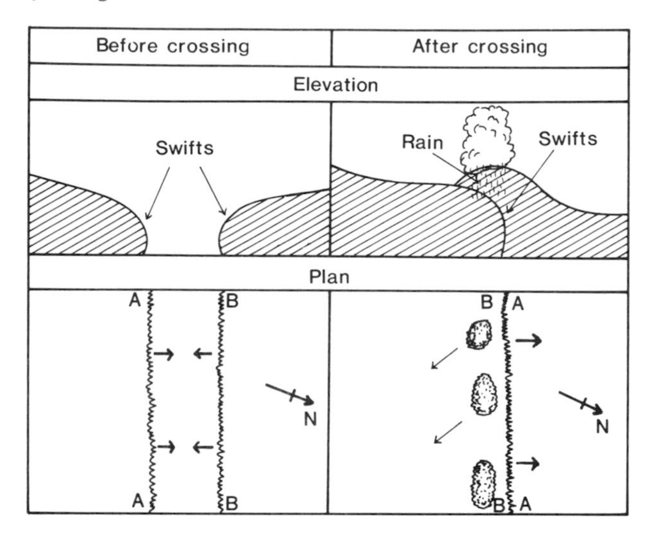

Fig. 15 Interpretation of radar echoes from crossing fronts, 5th July, 1966 (see text).
A–A Echoes from Swifts on gust front
B–B { Echoes from Swifts on sea breeze front (before crossing).
Echoes from rain showers (after crossing) (from Rider & Simpson 1968)[148].

Hirundines rely on insect food at lower levels than do Swifts, and have not been observed in large numbers in convergence zones. There is, however, evidence that the House Martin is a regular forager at high levels, perhaps at all times except when feeding young, rising to considerable heights on thermals and making use of insect concentrations. Hirundines are not uncommonly seen hawking low over the sea after a warm offshore drift has carried the insects from the land. During the evening descent of insect swarms, Swifts often fly low, and on these occasions suffer mortality when feeding over roads.

Black-headed Gulls, Starlings, and occasionally terns, take advantage of insect concentrations, and show dipping and twisting flight while feeding. Migrating Black Terns have been recorded hawking for insects over dry land on mild still days, but in cooler windier weather feed over a more normal aquatic habitat. Even pelagic species – Manx Shearwater and Sabine's Gull included – have been observed over coastal areas showing an apparent aerial feeding behaviour in warm weather.

WINTER FEEDING

In their winter quarters, Palearctic aerial insectivores, along with their local counterparts, find ample opportunity to feed in convergence zones. The movement and intensity of the ITCZ over the African continent (Chapter 1) determines the duration, amount and distribution of rainfall. The instability in this zone results in widespread thunderstorm activity, with areas of convergence concentrating insects in vast numbers at considerable heights. It is significant that in many areas of Africa, Swifts are seen in large numbers (totalling thousands) only in association with the gusty winds of thunderstorms. Squall lines are particularly frequent in west Africa during the rainy season. The Swifts remain for the duration of this weather and are hardly ever seen in clear skies. The distribution of Swifts in Africa probably mirrors the position of the ITCZ, since it is the rain associated with the thunderstorms that is responsible for the hatching of the myriads of insects that make up the food supply.[107] Insects are then borne on the wind systems of the ITCZ, often to be carried well away from the rainy region. On passage, huge numbers of Swifts have been recorded over the Sahel in west Africa in August – coincident with the arrival of the ITCZ at its northern limit. Hobbies and a few other raptors are largely insectivorous in Africa, and show a similar distributional behaviour to Swifts. They feed in hundreds on flying termites, which emerge after thunderstorms, but are rarely observed in dry weather.

I mentioned earlier the ability of aerial insectivores to recognise the situations in which their food becomes abundant, and in this respect, species in Africa may be able to show this ability at much greater ranges. Hobbies move from rain area to rain area, and detect them from up to 160 km distant. Many species can home rapidly to grass fires to feed on insects, presumably detecting from some distance either the smoke plume or the Cumulus-capped thermal formed by the intense heat.

POOR WEATHER

The presence of rain thus has a strong positive correlation with insect emergence and distribution in the tropics, but has the opposite effect when temperature levels are much lower. Cold, wet or windy weather during the local summer in northern Europe or southern Africa is notorious for causing great stress and occasionally considerable mortality to swift and hirundine populations. It has the effect of reducing the numbers of airborne insects except perhaps where they are newly emerging. There is often a concentration of aerial insectivores at low levels in such weather, when both Swifts and hirundines can be observed hawking close to water surfaces. Weakened Swifts often become prey to Hobbies that cannot themselves find sufficient insect food.[88] The foraging rate of Purple Martins in Texas has been correlated with temperature; there appears to be ample food above 13°C, but below 6°C foraging is unsuccessful. Such temperature thresholds naturally vary from region to region with species of bird and insect, but this illustrates the dependence of aerial insectivores upon temperature. The variation in behaviour that feeding birds adopt when food is scarce is mainly concerned with locating areas which are more sheltered from the elements and therefore perhaps harbouring a higher number of airborne insects. Hirundines hunt close to shelter where their high manoeuverability stands them in good stead. They will even fly in heavy rain if there is enough feeding close to trees and buildings.

David Lack[88] described Swifts avoiding thunderstorms and frontal rain, although he did emphasise the differences between northern latitudes and the tropics, where Swifts are associated with thunderstorms. He did not know the precise nature of the feeding behaviour in convergence zones ahead of thunderstorms in Europe, but it is similar to that in the tropics. Nevertheless, Swifts do tend to avoid the precipitation regions of these storms. Warm season thunderstorms usually develop in airmasses where insect activity is high previous to the rain, but in frontal situations this may not be so, with a lack of upcurrents to concentrate insects that *are* airborne. To overcome this, some members of the swift family have developed a rather unique behaviour.

On the approach of a frontal depression, Swifts remaining in their home area will be subject to more prolonged rain than if they fly away from the track of the depression, and therefore through the rain to its rear edge. Studies of the Swift in Scandinavia and England,[82,83,88] and of the Black Swift in British Columbia,[176] have revealed that vast numbers – in some cases several thousands – frequently leave breeding areas to avoid such weather. They tend to fly against the wind on the left hand flank of an approaching depression, where on average the most prolonged rain occurs; this manoeuvre takes them into a less rainy area. M. Udvardy[176] suggested that they might enter the warm sector where the warmer air may hold more aerial plankton. Many of the depressions are occluded, however, and it is more likely that the unstable air to the rear of the depression will provide the convergence necessary for optimum feeding, especially as summer temperatures enhance thermal activity. The most unstable air is found to the southwest of the track of an occluded eastward-moving depression (where it will have been swept in the circulation) and so the Swifts are moving into the best feeding conditions by the quickest route. It is probable that it is the first-year non-breeders that are largely involved – gathered from a wide area – since many breeding pairs remain in the nest holes in poor weather. The return movement in fine weather may take place at altitudes exceeding 3 km,[204] with ringing recoveries indicating complete journeys of up to 1,600 km. Scandinavian Swifts apparently make more massive weather movements than British birds, and Lack[88] suggested that this was due to the more rapid passage of depressions over Britain. There is certainly a tendency for these to become slow moving over northern Europe. Movements may be initiated by the early signs of an approaching depression, for the Swifts begin to leave well in advance of the rain. Some large summer movements in Britain are not always associated with depressions, but nonetheless occur in overcast weather with moderate to strong winds.[82]

Figure 16 illustrates the synoptic situation which gave rise to a massive southerly movement of Swifts in eastern England.[83] Large numbers of Swifts

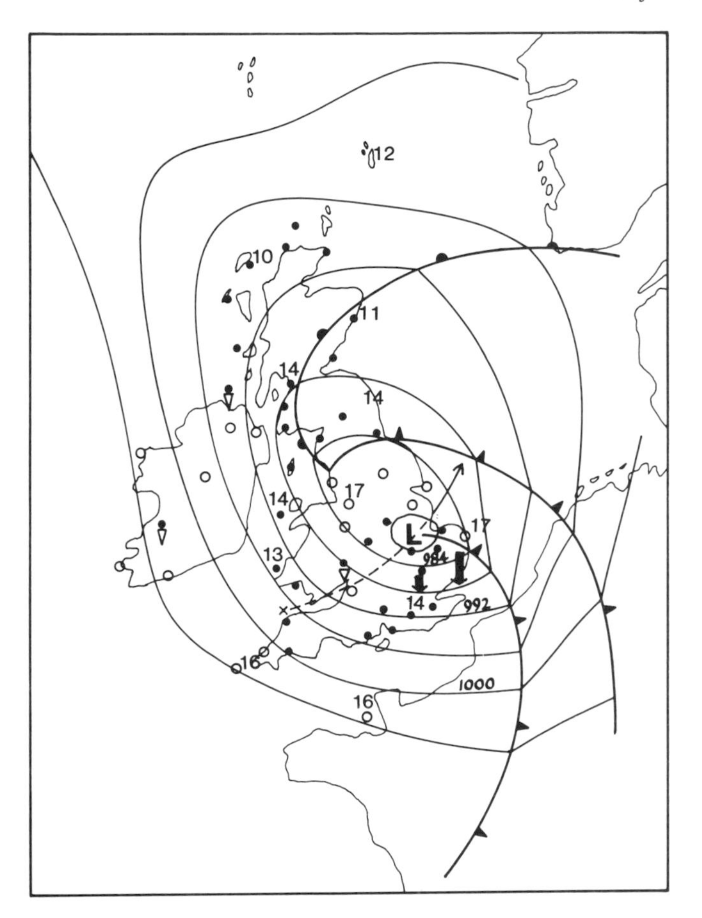

Fig. 16 Swift weather movements, 29th July, 1956, 1200 hours. ➡ Swift movements at chart time. Movement of depression shown, × position at 0600 hours. Temperatures in °C. ○ No rain ● Rain ▽ Showers.

had already been moving the previous day from a thundery depression over northern England, and huge concentrations were observed on the afternoon of the 29th July, 1956. It is difficult to see how the energy used by the birds in making some of these upwind flights can always be compensated by the feeding undertaken once the outward movement is terminated. In this instance, however, the unstable airmass to the rear of the depression gave ideal feeding conditions the following day, with high surface temperatures initiating marked convection. In cases such as this, the Swifts can then follow the improved feeding back to the breeding areas. Lack stated that the autumn migration begins at the end of July, but that it is not initiated in such weather conditions. Nevertheless, there is a high probability that many of the

birds in the July 1956 movement did not return, particularly the non-breeders.

In a wet summer, the Swift needs to remain in the breeding area for a longer period after the young have left the nest, in order to store fat for migration. Those with late broods, however, must leave immediately the young have flown, although the inability to accumulate sufficient fat reserves may result in a failure to migrate, and subsequent death.[88] Hirundines may abandon their young under these circumstances.

At the beginning of the breeding season, Swifts stay away from their nest holes for long periods during cold weather. Otherwise the diurnal cycle of insect activity regulates foraging, so that they are less often at the nest between noon and 1600 hours during which time airborne insects are most numerous. (The feeding activity of the Hobby shows a similar variation.) Lack[88] showed that each meal brought to the nestling Swift has the same mean weight in all weathers, but each adult must spend a proportionately longer time gathering the insects under poor feeding conditions. In fine weather, large insects of between 5 and 8 mm in length are caught, although still outnumbered by smaller ones of 2 to 5 mm, but when food is scarce, a Swift cannot afford to be selective – taking a greater proportion of small insects. Lack recorded an instance in which Swifts brought much food to the young in continuous rain, comprising insects normally only active in warm sunny weather. It subsequently materialised that they had flown through the rain belt to feed in clear weather beyond, and followed this (presumably a frontal clearance) as it moved towards the nest site. Although only a short distance, this movement was analagous to long distance weather movements.

With no similar capacity to avoid poor weather (though both House Martins and Sand Martins are sometimes associated with the weather movements of Swifts) hirundines occasionally suffer quite dramatic losses. This is particularly evident on migration and in winter quarters. Torrential rain along the ITCZ in central Africa has killed large numbers of Swallows and Sand Martins, and in November 1968, a spell of very cold wet weather in southern Africa killed many thousands of wintering Swallows, including some from the British breeding population. The cold weather affected a very large area from Zimbabwe southwards. Few insects were on the wing after three days of poor conditions, and the wintering Swallows – along with House Martins – appeared to be prone to starvation at temperatures at which the local hirundines survived.[107] A similar event occurred in December 1970, when weakened birds invaded houses for shelter and warmth. Other examples of mortality during migration are given in Chapter 8. Abnormal rainfall has also assisted temporary expansion of wintering range. In March 1961, phenomenal rains in parts of South Africa allowed insect and animal life to flourish in a normally poorly vegetated and arid area. Indeed, the irregular droughts and rains of southern Africa have profound temporary effects upon wintering hirundines.

WEIGHT VARIATION

Measurements of the weight of aerial insectivores have demonstrated just how much variation there is with regard to the weather and its effects on food resources. The meteorological element most closely correlated with weight is temperature. It is, of course, an indirect correlation since the temperature and, to a lesser extent, rainfall affect the airborne insect population (as shown earlier) and thus the weight of an individual bird.

In a study of Swift weights in an English summer,[56] a marked reduction was found to take place in cold weather. The average weight of birds trapped on days with a maximum temperature below 16°C was 39·5 g, while on warmer days the average was 42·8 g – a figure close to the normal for nesting adults. There were also marked fluctuations in the weight of individuals trapped more than once on different days. A suggestion that breeding stresses may have been a prime influence on weight changes was overruled, and later supported in a study of another aerial insectivore. A similar correlation between weight and temperature was discovered in a hirundine, the Crag Martin, wintering in Gibraltar, where there are no stresses other than those of finding enough food and shelter.[45] Figure 17 shows marked drops in mean weights during cold and wet spells; highest weights occurred after mild days of sunny dry weather, with light winds, i.e. when food was abundant. In strong winds, heavy rain and low temperatures – chiefly on the passage of cold fronts or depressions – insects were in short supply and, owing to poor light, the birds were unable to locate food efficiently. This resulted in severely curtailed feeding times, especially in midwinter, compounding weight losses due to food shortages. A substantial proportion of weight was

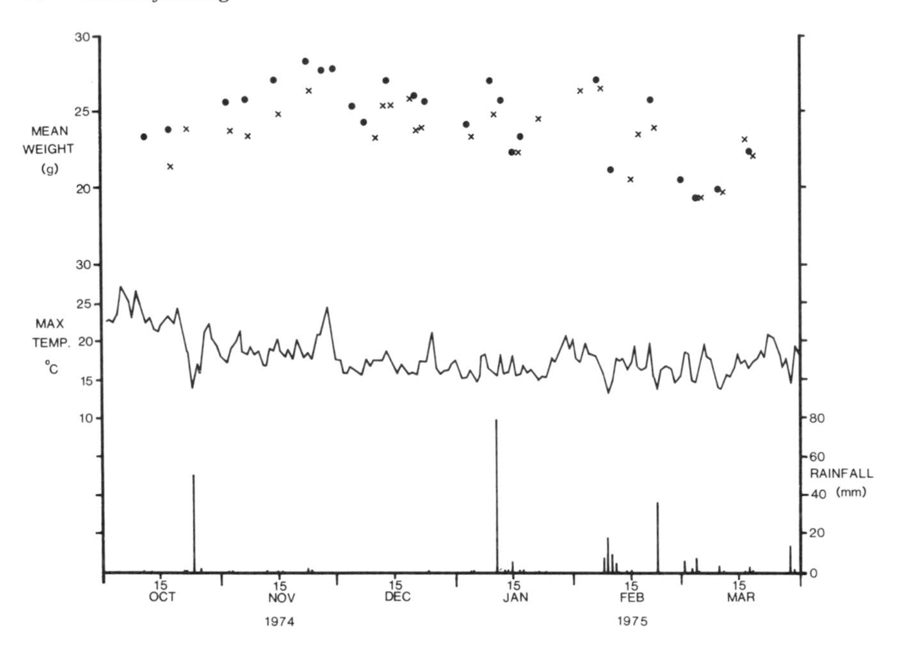

Fig. 17 Fluctuations in the weights of Crag Martins wintering in Gibraltar, October 1974 to March 1975.
● Mean weights of evening samples newly arrived at the roost
× Mean weights of morning samples leaving the roost
The build-up of fat reserves in early winter is clear, as is overnight weight loss. Weight losses after cold or wet spells are soon compensated by gains in warmer drier weather. Mean weights in October, November and March may be influenced by the presence of passage migrants. Most rainfall was associated with the passage of cold fronts.

lost in periods of cold and/or wet weather – some individuals losing up to 25% in a few days. Mortality was observed at the roost, probably as a result of starvation. There was no evidence that the survival rate of small light birds was significantly different to that of the larger heavier individuals.

SUMMARY

The topic of this chapter has been the behaviour of species feeding on aerial plankton. Flying insects and tiny arthropods are concentrated in the atmosphere by zones of convergence. These may occur simply in thermals, or in more complex small-scale features such as sea breeze fronts and thunderstorm squall lines. The formation of each has been described and reference made to the feeding habits of aerial insectivores in breeding and winter quarters, where in Africa, birds make use of the ITCZ and its manifestations.

A breakdown of the warm airflows which stimulate and concentrate aerial plankton causes considerable stress in these species. In cold wet weather, the scarcity of food may initiate massive long distance weather movements in certain species of swift, and cause large-scale mortality to hirundines. During these periods, the weights of such birds fluctuate markedly with temperature.

CHAPTER FIVE

Breeding

It is generally accepted that for the majority of species the ultimate reason for breeding at a particular time is to ensure that the young are hatched when food is most abundant. The time of breeding of most land birds in the seasonal climatic regimes of low latitudes ultimately depends upon the fluctuations in food resources controlled by the rainfall of the Inter-Tropical Convergence Zone (see Chapter 1). In mid latitudes, reproduction in a large number of species in spring broadly corresponds either with the maximum availability of soil invertebrates (which may retreat during soil drying conditions in summer) or of insects and defoliating caterpillars. Seed-eating species also breed at this time, since seeds are more plentiful in summer, and some must supplement their diet with invertebrate animal food during the breeding cycle. In high latitudes, where the breeding season is short, reproduction is dependent partly upon the spring thaw of snow and ice but, as in other zones, is often timed to take advantage of the maximum availability of food. In mid and high latitudes especially, the weather sometimes plays an important part in the timing of breeding, particularly temperature, in its role of controlling plant growth and hence animal food.

Once the reproductive cycle has begun, the degree of influence exerted by the weather depends to a large extent upon the life style of the individual species. There is proportionately more damage or destruction to nests and their contents in open situations than in, say, thick vegetation; and hole-nesting species are probably safest of all. The successful rearing of young demands an abundant food supply; when this is unpredictable there are breeding strategies that partially alleviate the risks and consequences of starvation.

DISPLAY AND COURTSHIP

The breeding season is heralded by song, display, courtship, pairing and nest site selection, all of which are stimulated by mild sunny weather but are delayed by cold, wet or snowy conditions. For migrants, some of these activities may begin in winter quarters. This holds especially for those species with the compressed breeding seasons of high latitudes. Many wildfowl pair

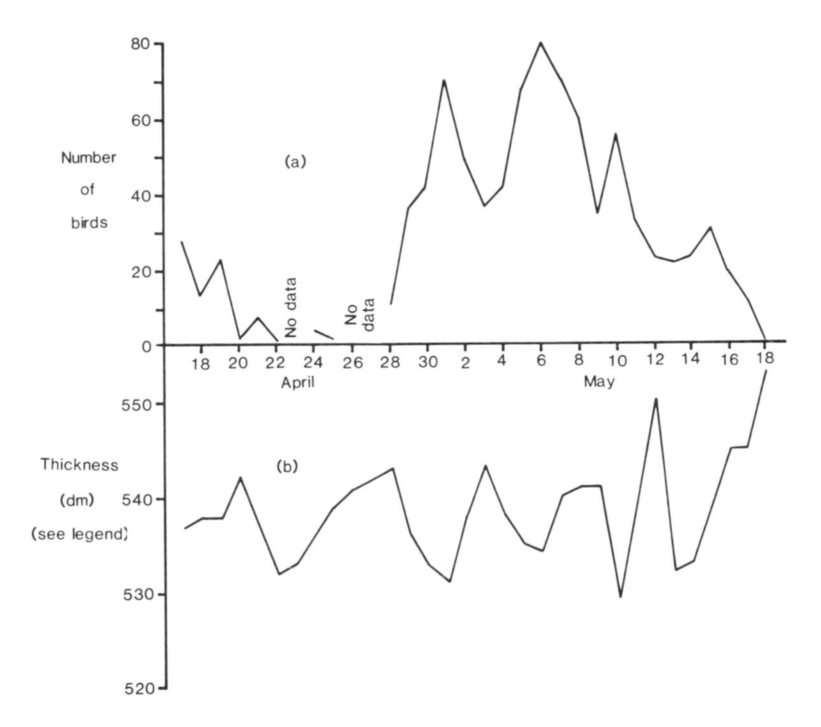

Fig. 18 (a) Number of Long-tailed Duck on Loch Branahuie, Isle of Lewis, Scotland in late spring.
(b) Type of airmass present, as shown by values of the 1000 mbar to 500 mbar thickness at Stornoway (in decametres).
Thickness values represent the depth of atmosphere between the 1000 mbar level (near the earth's surface) and the 500 mbar level (at an altitude of approx. 5·4 km). Since cold air is denser than warm, low values indicate a polar airmass while high values show a tropical airmass (from Elkins 1965)[42].

early to allow an immediate start to nesting on reaching their breeding grounds. For example, in the Outer Hebrides late spring movements of wintering Long-tailed Ducks from the sea to coastal lochs have been correlated with airmass type[42] (Fig. 18). The movements were assumed to be territorial in nature, and invariably occurred three days after an incursion of mild air which probably stimulated the birds physiologically to visit a habitat resembling that used in summer.

For some resident birds, sexual behaviour is shown in the depths of winter. Fine weather stimulates grouse and Ptarmigan to re-occupy territories and display, but in unfavourable weather birds remain in often large flocks for weeks or even months. Display, pairing and territorial behaviour begin suddenly if a long snowy period terminates in a sudden thaw and calm weather. In Sweden, lek activity of Black Grouse may be inhibited by a smooth and featureless snow surface, and display may even take place in trees if the snow is loose and deeper than 6 cm.

Raptors normally display only in fine weather. Resident eagles, including the Golden Eagle, display on sunny days, most often in late winter and early spring. Much of the display flight of raptors is undertaken in upcurrents – both thermal and orographic (see Chapter 2). Favoured areas for these currents are topographical features such as peaks and ridges – often attracting birds from several territories. Sparrowhawks, kites, buzzards and harriers use updraughts in high level nuptial displays[20] – some species congregating in small parties – and the Short-toed Eagle exhibits its solitary 'sky-dance' display only in strong upcurrents. Woodland raptors display over areas where thermals are weak; nevertheless, upcurrents *are* present on fine days. The crowns of trees, when in leaf, absorb enough radiation to raise their temperatures to a considerable degree[54] – acting as a thermal source. The thermals may not be very powerful or sustained, and are later in forming than the stronger currents over open country, but are probably sufficient for a displaying bird to use. Additionally on windy days, considerable lift is generated over upwind woodland edges, where raptors may be joined by displaying corvids.

SONG

Song carries further during certain meteorological conditions, though the frequency of the song, rather than the audible range, is possibly of greater biological importance. The acoustics of natural surroundings change with the seasons and the weather. Audibility of sound is low with the steep temperature lapse rates associated with instability, and in turbulence created by strong winds, when it is also suppressed by the noise of waving branches and foliage. Audibility is high when an inversion is present at low levels, since the sound waves are then reflected downwards to some extent. This may account partly for the high level of song early on a fine calm morning when an

inversion is most likely to exist, and turbulence is minimal. Audibility is greater at night than by day for these reasons. It has been found that bird sounds in the early morning can be as much as 20 times more effective in the area of coverage than those made at midday.[70] The dawn chorus of song birds is a familiar phenomenon in fine weather, and wind tends to delay the first morning song. A study of the midwinter song of the Song Thrush revealed an increase in vocal activity in mild weather (territorial song stimulated) and also when sharp ground frosts occurred (with an accompanying inversion).

Most treetop songsters, including the Woodpigeon, Coal Tit, Song Thrush, Blackcap, Chiffchaff and perhaps Goldcrest, sing less in strong winds, but there are exceptions – the early breeding Mistle Thrush springs to mind, and its name 'Storm cock' is not inappropriate. One of its more eastern relatives, the Red-flanked Bluetail, is also reputed to be a very persistent songster in rain and wind in Japan. It may be the rise in temperature that stimulates their song, since mild weather early in the year is often accompanied by windy and wet conditions in maritime climates.

The Skylark sings more frequently from the ground in very poor visibility and in strong winds, and the height of its song flight may change with the intensity of thermal upcurrents. Suzuki and others[166] found that Skylarks in Japan sang at a height of 40 to 70 m when the surface air temperature was 16°C, but when the temperature rose to 24–28°C their song was delivered from heights of 80 to 100 m. In a cooler maritime environment, flight rarely exceeded 20 m in height.

Other workers have related song variation to factors such as light intensity and relative humidity, each of which doubtless plays a part, although the latter may be merely concomitant to other factors. Damp cloudy days with

drizzle are typically associated with the 'reeling' of the Grasshopper Warbler and the drumming display of the Snipe. It may be that both these rather unique sounds carry better in such conditions. Certainly this type of weather is most often associated with high humidity, low cloud and stability in the lower layers of the atmosphere, particularly in the evening, a favourite time of day for Snipe display. Displaying Black Grouse call more loudly and persistently in fog to compensate for the reduced visibility, but if the fog is dense, display may be weaker. Winter lekking occurs most frequently in anticyclonic weather, when audibility is greater.

ARRIVAL ON BREEDING GROUNDS

Though resident species may be able to breed as early as conditions allow, the timing of breeding is nowhere more vital than at high latitudes where the season is short. In the Arctic, all fresh water is frozen until at least May. The land is often completely snow covered, although in arid regions the ground may be bare, especially where exposed to strong winds. Ice is extensive off seashores which do not experience warm water currents. Leads between the sea ice and the fast ice along shores are opened by outbreaks of mild air in spring. Overland, the thaw depends upon the exposure and aspect of a particular site, and the extent and depth of the snow. Timing of the thaw varies according to the position and track of travelling depressions and the extent of their warm surface airmasses. Rivers are the first fresh water bodies to thaw, followed by ponds and lakes.

A late thaw can prevent successful breeding and the length of the period from laying to fledging is vital in determining the northward limit of a species' range. For example, the Whooper Swan takes 130 days to complete its breeding cycle. Within that part of its breeding range north of the Arctic Circle, only half the summers are ice-free for this length of time, and at these latitudes breeding is often unsuccessful. Its smaller relative, the Bewick's Swan, can breed further north, since it has a cycle of only 100 to 110 days. On arrival swans and geese begin to nest immediately on patches of higher ground which are snow-free. These sites have the added advantage of lying above the level of any thaw floodwater.

The problem of finding sufficient food on arrival at the breeding grounds is one that confronts all species in high latitudes. Water birds await the inland thaw on open leads in the offshore ice. A late thaw depresses water temperatures, thereby increasing maintenance needs and delaying breeding. On arrival at the breeding site, insectivorous species such as waders and passerines initially take tundra plants, the previous years berries and seeds and other vegetable matter until insects become plentiful, although deep snow may render even these food sources unobtainable.

The breeding of waders in many arctic regions is timed to take advantage of the greatest abundance of insect food in July. In northeast Greenland, however, breeding is directly related to snow conditions.[63] The difference between the two strategies lies in the duration of food availability. In the latter region sufficient food is obtainable over a long period and the birds time their breeding to coincide with the snow melt. The date of snow melt, and thus the availability of a nest site, naturally affects laying dates, but although a late thaw may preclude breeding, variations in snow cover, time of thaw and local food abundance from valley to valley nevertheless ensure that there is at least partial breeding success within the population.

If on arrival the weather is poor or the thaw is late, it is an advantage for migrants to have some degree of fat reserve to enable them to cope successfully with the food scarcity. For example, Turnstones are well adapted for survival, arriving with reserves of fat. In arctic Canada, they take any food available if extensive snow is still present; Macdonald & Parmelee[95] watched a flock collecting overwintering insect larvae from beneath mud flakes in temperatures of −10°C accompanied by strong winds, and in such circumstances there is no competition from other species. Nevertheless, mortality from starvation does occur, particularly if poor weather on migration has already depleted fat reserves.

In less severe climates, late snow or frost can cause great problems when it falls at a critical time, since the breeding birds are less prepared.[130] It has been suggested that apparent differences in the degree of mortality among species are due either to variations in the fitness of individuals (via contrasting fat reserves) or to subtle variations in the timing of their arrivals.[184] In Scotland, mortality to Oystercatchers has arisen in such weather after they have returned to inland breeding sites. In late March 1979, 23 cm of snow fell in Deeside and Adam Watson found 33 dead birds on 50 km of road.[181] The birds were emaciated, with very little food in their gizzards. Higher up the same valley, where severe drifting had kept half the grassy areas snow free, sufficient food was available, and no dead birds were found on 30 km of roads. Watson thought that a retreat towards the coast had been precluded by the proximity of the breeding season, and that the mortality (during a period in which local Lapwings had suffered little) pointed to the fact that Oystercatchers are not well adapted to snow – inland breeding being a relatively recent phenomenon. It is of interest that, in Britain, adult Oystercatchers add weight throughout winter to cope with the early return to the breeding grounds. This is in contrast to immature birds, which are much later in arriving, and to other northern waders, which gain weight before departure and at staging posts en route.

Just as in the arctic, snowfall in upland Britain determines both timing and success in the reproduction of certain species, and is often critical. In Scotland, Golden Plover, Meadow Pipits and Wheatears return to their high level breeding grounds once there are snow-free patches. Some plovers arrive even when there is still a considerable amount of deep snow, but there are fluctuations in numbers as flocks move to lower levels in periods of cold or snowy weather. After mild winters they may arrive as early as February, but after a snowy season breeding may not begin until after mid May. In some years snow remains deep on high ridges, with snowfall still possible in July. Nethersole-Thompson[116] records exceptional May snows when 'trips' of Dotterel are still to be found – flocks which have been unable to take up individual territories. Normally Dotterel arrive in early May when the first snow-free ground appears above the 900 m contour, but deep snow may delay their return and force some birds to nest in other lower and more southern areas.

THE NEST

The time taken by a bird to build its nest is controlled by a variety of factors, not least of which is the time needed to forage. Temperature therefore has an indirect effect on nest-building, but the two more important elements involved are wind and rain.

Strong winds tend to inhibit nest building in exposed sites. Since Swifts catch wind-blown pieces of grass and other flotsam for their nests, building is irregular, being most frequent when the wind is strong enough to lift material into the air. Strong and gusty winds also affect visits to the nest by making flight less accurate and misjudgements at the nest entrance more frequent.[88]

The structure and material of a nest may depend upon humidity and rainfall. There are several species for which wet nest material is important. In abnormally dry springs the smooth lining of woodpulp or dung of a Song Thrush's nest (and even the mud strengthening) may be missing, leaving the nest as a soft structure lined with just twigs and grasses.[156] The Wren is also dependent upon sufficient rainfall, since the compactness of its nest is achieved by the use of wet material which shrinks to a tight structure when dry. There is a possibility of disintegration if the nest is built in dry weather. The Rev. Edward Armstrong[5] found that building often began at the onset of a shower, and listed other species which need similar building conditions, e.g. the Chiffchaff, Marsh Wren, Great Reed Warbler and Dipper. However, Reed Warblers cease to build when rain falls, so that a period of fairly continuous wet weather might prolong building considerably. Owing to the frailty of the basal platform and its consequent likelihood of disintegration in rain and wind, the female works longest and hardest in the early stages of building.

The function of a nest is to protect the vulnerable eggs and young from predators and adverse environmental conditions. Naturally the influence of the weather on the completed nest and its contents varies in importance according to the nest structure, site and exposure. A nest site may be chosen or adapted to minimise adverse meteorological effects. Birds nesting in open sites often ensure that the nest is sheltered from, or facing away from, prevailing strong winds, or that the maximum heat from the afternoon sun does not fall directly on eggs and young. Nevertheless, exposure to the sun may be beneficial in cooler climates, and in higher latitudes southern exposures prevail. There are exceptions to this rule, of course. Some desert species orientate the nest to avoid the prevailing wind early in the season when it is cool, in order to reduce heat loss. Later, when the heat is more intense, nests are built facing the wind to enhance evaporative cooling of the young. Some alpine birds also show a seasonal change in nest exposure.

Nests built in unstable vegetation (particularly reedbeds) are the most likely to be damaged or destroyed by wind and rain. Those in tall trees are at risk in strong winds, especially of the Osprey, which often builds in exposed crowns; in one North American study, 7% of 203 nests were destroyed by wind. Osprey pairs which have lost their nest do not re-lay, but sublimate their breeding drive by building a 'frustration' eyrie. The increasing and now well-established Scottish population suffers occasionally from gales and, in July 1978, five young were lost when two nests blew down. Occasionally the nesting tree itself is damaged or felled.

Rooks' nests are more vulnerable to wind than those of other corvids because of their greater exposure, although cliffnesting Ravens have had nests blown from their foundations by very strong updraughts. In Cornwall, where strong winds are relatively frequent, half the Rooks' nests are built in evergreen trees, chiefly Scots, Monterey and Maritime Pines. In two well-sheltered rookeries, about 50% more nests survived a ten-month period in evergreens than in deciduous trees. In many other less windy parts of England, the majority of nests are built in deciduous trees. In Scotland, rookeries are more often found in evergreens, but situations may be related more to the distribution of tree species rather than a matter of shelter, since in the very windy district of Caithness, less than 15% are in Scots Pine and more than 50% in Sycamore. The shelter afforded by evergreen foliage also varies with species of tree and height.

Birds nesting close to open water, on unstable banks or in sites with a high water table or non-porous surface are vulnerable to heavy rain and flooding. Flood risk is greater if birds have nested low during a spell of dry weather. The Little Ringed Plover sometimes nests on flat sites exposed by desiccation in spring, which may then be prone to flooding in heavy summer rains. Parrinder[132,133] found that in some years rainfall was a significant factor in the loss of nests.

In regions subject to regular summer drought, wetland birds are influenced very much by water levels to the extent that populations and breeding sites fluctuate markedly. Small shallow pools dry up in seasons with low rainfall and newly-flooded areas are readily colonised. After dry winters, the breeding of a number of species in Mediterranean regions is delayed or not undertaken at all. Conversely, wet winters and springs provide optimum conditions and populations increase. Examples in Britain are the failure of the small Black-tailed Godwit population to breed after a drought, and fluctuations of Garganey according to the presence or absence of shallow pools. Indeed, breeding populations of other ducks vary annually according to spring flooding.

The construction of a water bird's nest may affect its vulnerability in rising water. Floating nests, such as that built by the Great Crested Grebe, are less vulnerable than those anchored firmly (for example, the Pochard and Moorhen). Waves generated by strong winds on larger bodies of water also do considerable damage. On a 46-hectare Hertfordshire reservoir, wave action and heavy rain were responsible for more loss of Coots' eggs than any other factor, amounting to 33% of the causes of breeding failure over a ten year period.[150] This factor varies in importance according to the topography, since other breeding studies of the Coot have resulted in very contrasting figures. The Moorhen nests in less open situations than the Coot, with more protection from vegetation. Nevertheless, losses from flooding gave a failure rate of 13·7% for 1,154 nests in Britain. Freethy[52] recorded an amazing joint effort by a family of Moorhens (two adults and three almost fully-grown young) in building another brood nest for the second brood of four in danger of being swamped by rising water. Within fifteen minutes of the level beginning to rise, the second brood was safely transferred, while the original nest had sunk beneath the water. After two days the water had subsided and the four young were back in the first nest.

DATE OF LAYING

It is widely recognised that daylength and temperature are important proximate factors governing the precise date of laying. Each species is physiologically adapted to a range of temperatures, and low temperature tends to inhibit breeding by retarding sexual processes.[101] The relationship of temperature to the date of laying is, in most cases, indirect. Increasing warmth in spring may be a sign to an individual that food will become more plentiful, or that less food will be required for maintenance needs and therefore more will be available for breeding activities. Since both summer migrants and residents show this link between temperature and laying date, it would seem to be the effect on the environment that is important, rather than the level of temperature itself, as the summer visitors are not present during the more marked temperature fluctuations of early spring.[87]

There is a tendency for resident species to breed earlier when the preceding winter has been mild, as this improves the body condition of the individual. Winter conditions in mid and high latitudes often have a profound effect upon the success of the future breeding season for the survivors. If the weather is severe for a prolonged period during winter or early spring, and food becomes scarce, increased maintenance needs result in a reduction of the energy available to the female for egg production, and the number and viability of the eggs laid in the spring may be lower.

In arctic North America, Brent and Canada Geese form eggs after arrival in spring, thus being able to time laying according to local conditions. In contrast, the Lesser Snow and Ross's Geese nest as early as possible – beginning egg formation before arrival. The former two species fail to breed in a late thaw, and the gonads regress if no habitat is available, while the latter two breed but may lose nest and eggs under such adverse conditions. The return itself may even be delayed by the inability to accumulate fat for the migration. For example, there is evidence that goose migration from western Scotland is later in cold springs because of the suppression of the growth of new grass.[189] Since the first phase of the breeding cycle of arctic geese is undertaken in conditions of relative food scarcity, fat and protein reserves existing on arrival must sustain them for a considerable period – often up to three weeks. Indeed, adverse migration weather weakens many high latitude species so that they arrive in too poor a physical state to be able to breed successfully.

The importance of a food supply independent of current environmental conditions is illustrated by the breeding habits of those northern passerines whose food is controlled in part by climatic factors during the previous summer. Northern forest trees normally have a fruiting cycle of two or more years, so that there is a more or less abundant supply of seeds every other winter. In other years, the tree is 'tired', and produces few seeds. This cycle can be altered by the weather during the flowering stage. Frosts may kill the flowers so that the cycle is interrupted, but high temperatures may assist flowering and fruiting, enhancing the supply of seeds in years when the tree is productive. Amongst others, Mealy Redpolls, Siskins and crossbills all feed on conifer seeds, and breed at the peak of cone crop production, which varies with area and tree species.[121] If food is plentiful, Mealy Redpolls nest up to two months earlier than normal, and these, crossbills and Nutcrackers are known to nest in March and April with temperatures down to $-20°C$ and snow as deep as 50 cm. The Parrot Crossbill times its breeding so that hatching coincides with the opening of hard pine cones in the sun's warmth, thus ensuring that the seeds are easier to obtain. In Scotland, there may be exceptionally early breeding of crossbills and Siskins in years with heavy pine or spruce cone crops.[118]

In several species the date of laying is determined, in part, by conditions immediately preceding the event rather than by those predicted for hatching time. This strategy is a modification to suit particular conditions. Certain population studies suggest that for insectivores (including some waders) there is a correlation between egg-laying and the appearance of insect food sufficient for the egg-forming female. There is evidence that this strategy may be followed for many species, but only a few have been studied in enough detail for this to be clearly shown.[87,136]

The onset of laying in the British populations of Great and Blue Tits depends to a large extent on the availability of defoliating caterpillars. These, in turn, are dependent on the effect of early spring temperatures upon the appearance of leaves, and diminish quickly in abundance later in summer. The female Great Tit cannot lay until there is enough food, but must ensure that its young hatch before the food supply deteriorates; thus it steers a middle course. It must be emphasised that this strategy is not necessarily essential in other regions where the type and timing of the staple food differ. If Great Tits have begun to lay in fine weather, and there follows a sudden drop in temperature, only those individuals which are laying, or have started to form eggs, will continue to lay, but no others will do so until temperatures recover. Individuals begin to lay about four days after the onset of warm weather, and this delay is related to the activity of the caterpillars. Birds which have already begun laying or forming eggs need rather less food than those which have not yet begun.[137] That maintenance needs play a significant part in egg laying was shown by studies in which female Great and Blue Tits roosting in warm nest boxes laid earlier than those roosting in cool boxes; they lost less weight overnight, and therefore had energy reserves to form eggs sooner. In relation to the completion of the nest, tits lay later in a mild spring – perhaps by 8 to 12 days – than in a cold spring, when eggs may be laid before the nest is lined. Exceptionally, eggs may be laid only three days after building has begun.

The Swift shows a similar strategy to that of the Great Tit, since its future food supplies are unpredictable. The time of laying is determined by the current insect abundance, and the clutch is begun about 5 days after a marked rise in mean temperature.[88]

While, as we saw earlier, a few coniferous forest species breed at the time of a peak food supply which is unrelated to the immediate weather, most birds refrain from nesting until spring. Just as mild weather stimulates territorial behaviour and courtship, the continuance of such conditions will finally result in nest building and laying and, in extreme situations, nesting

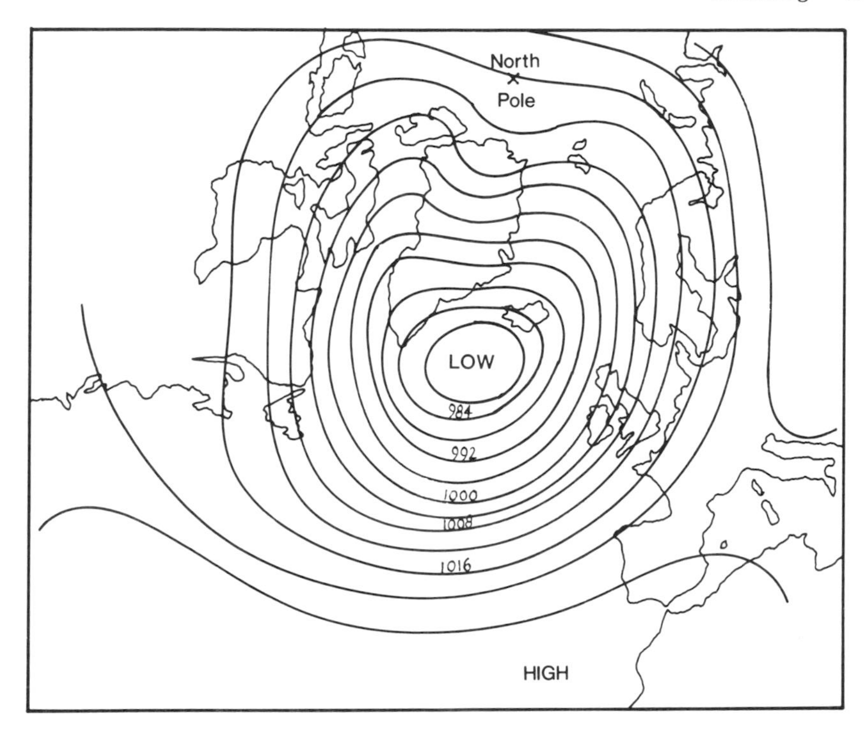

Fig. 19 Mean sea level pressure, January 1974. Early breeding in Britain. At least twelve warm sectors affected southern Britain during the month in the circulation of unusually vigorous depressions. Compare with normal mean January pressure distribution (Fig. 47).

and successful rearing of young may take place in the depths of the northern winter.

Probably the most notable winter breeding in Britain occurred in the season of 1953/54,[161] when there was an abnormally long period of warm weather during November and the first half of December. The two months were characterised by strong SW winds, with the mean pressure in the Iceland region 13 mbar below normal and that over southeast Europe 12 mbar above normal. This had the effect of markedly tightening the pressure gradient over the British Isles. The mean temperature between mid November and mid December for England and Wales was 5·6°C above normal, and at Oxford it was the mildest November and December for over 100 years. Many plants flowered and the warmth stimulated an outburst of breeding in several species. Successful fledging was known to have taken place in one Blackbird and three Song Thrush broods. The two successful thrushes in that winter are frequently the commonest species to breed out of season in Britain. One remarkable pair of Blackbirds at Oxford during 1961 laid the first egg around 23rd January, and went on to raise five broods between that date and July. This unprecedented breeding record was assisted by the prolongation of mild weather into February and March. In very mild winters the weather is controlled by unusually deep depressions in the circulation of which run frequent warm sectors across Britain in very strong southwesterly airflows (Fig. 19).

Early breeding is, of course, carried out by only a very small proportion of a population, and often permits the rearing of more than the normal number of broods, but should adverse weather intervene, then there is a strong likelihood of desertion. In the late 1960s and early 1970s, there was a tendency in England for cold springs to follow mild winters. Early breeding is likely to be unsuccessful in these circumstances, and successful nesting then follows later than normal.

The spread of laying of first clutches is often dependent on fluctuations in the weather. Laying is prolonged in a poor spring – only a few clutches being started at any one time. In generally fine weather, most clutches are started within a short period, but should fine weather rapidly succeed a period of poor weather, all laying is achieved within a very short time.

THE CLUTCH

In many species the number of eggs in the first clutch is partly dependent upon the food supply available before and during laying.[87] One theory postulates that clutch size in many European birds increases pole-wards in mid and high latitudes. Increased daylength allows a longer foraging time in higher latitudes, and thus a larger brood can be reared. This presupposes a food availability unchanging with space and time, but so often this is not so, and temperature (as well as other meteorological elements) becomes a proximate factor in the regulation of clutch size through its effect upon food. The same applies to those species which breed through a wide range of altitudes, such as the Meadow Pipit in Britain, whose clutch size diminishes with height above sea level. It is understandable therefore that, in periods of climatic variability, clutch sizes in any particular species may well differ within a region from year to year.

Population studies have shown that this is so, and that egg size, structure and laying intervals may also vary. During a food shortage, many species (especially aerial insectivores) suspend laying and limit incubation time and clutch size. David Lack[81] suggested that anomalous weather could mislead a bird into producing a clutch which is adapted to a season for which that weather is characteristic. The Swift lays three eggs when food is abundant, but in cold, wet or windy weather (or later in the season) it lays only one or two with a longer interval between eggs. Such regulation of the Swift's clutch size has great survival value in a poor season, with a lower success rate in broods of more than one. Variations in clutch size of arctic wildfowl can be attributed to the extent of snow cover, the rate of growth of food plants and the amount of energy used while awaiting the thaw. In some arctic geese the clutch size is reduced by one egg for every day the weather delays laying – so great is the need to lay over a short period of time.[128]

It is probable that slight differences in egg structure are invariably linked with environmental factors. The egg weight of the Great Tit can be correlated with the temperature on the day on which it began to form. In unusually cold weather, one female laid eggs which steadily decreased in weight until laying ceased after only six eggs.[137] After the exceptionally cold and wet April in 1973 (Fig. 29), several hen Greenshanks in northwest Scotland laid thin-shelled eggs, many of which broke in the nest. One bird laid a miniature egg containing no yolk and weighing only 9 g, 30% of normal. Pellets from the Greenshanks were analysed and found to contain fishbones, newt remains and rodent teeth. This diet perhaps results from a need to replace the calcium used in eggshell formation from a source alternative to the normal (but temporarily scarce) insect food.[117,119] Such a feeding strategy has been noted in the arctic, where S. F. MacLean[98] found teeth and bones of Lemmings in the stomachs of small sandpipers (Dunlin, Baird's, Semipalmated and Pectoral). He concluded that such calcium-rich food was necessary for efficient egg production in the severe climatic conditions of northern Alaska, and that variations in the availability of calcium possibly result in adjustments to egg-laying intervals – perhaps even limiting breeding populations.

The number of broods raised by a pair of birds is generally determined by the length of the breeding cycle in relation to the seasonal food supply. Only one brood may be possible in the short summers of high latitudes, but none is raised with a late thaw and poor weather. In temperate regions, some species lay further clutches to replace those lost for any reason (including weather) and, while certain pairs may lay several in mild and food-rich seasons, even double-brooded species only manage one if the weather is poor.

INCUBATION

The onset of incubation in relation to laying varies among species. With clutches of more than one egg, incubation beginning before the last is laid usually results in asynchronous hatching, which is an adaptation in species whose food resources fluctuate markedly. Arctic birds incubating continuously from the first egg ensure that not only does asynchronous hatching provide at least partial breeding success in an often hostile environment, but also that there is direct protection for the eggs from extreme chilling. With asynchronous hatching, the resulting differential in nestling size ensures the survival of only the larger, older, individuals when food is scarce. Thus precious food is not wasted.

The principal meteorological elements from which the clutch must be shielded are heavy precipitation, extreme heat and extreme cold. Rain may not be a significant problem to an embryo inside an egg, but it increases in importance when accompanied by low temperatures, or (if the eggs are exposed) wind, with subsequent cooling by evaporation. There is evidence also that the humidity of the environment plays an important role in hatching success of wildfowl eggs. Duck eggs hatch less successfully in dry summers if the habitat is also dry, and the proximity of water may be vital in that the duck can carry water to the eggs on its plumage.

Precipitation in the form of snow increases the risk of chilling the contents and of covering the nest. A greater amount of attentiveness by the adults naturally alleviates both risks but the date of laying and hatching in relation to the snowfall is critical. Losses of eggs are greatest in newly laid clutches, while losses of young are highest in the first few days after hatching.

Low temperature during incubation lengthens the time needed by a bird to cover its eggs and also the time necessary for foraging. If forced to leave its eggs for long periods in order to feed, there is a likelihood of desertion, particularly in the early stages. Kendeigh[78] found that as the daily mean air temperature rose, there was an increase in the total time a small bird spent off its eggs, but no significant change in the average length of the inattentive periods i.e. incubation stints became shorter but more frequent.

For birds in which the hen carries out all the incubation, her initial body condition may be crucial. A reserve of fat assists a hen bird of prey during any failure of the male to provide sufficient food (for example, in poor weather) since she will not need to hunt and therefore leave her eggs to chill.[122] However, delays and interruptions in incubation of up to a week can be tolerated by the embryos of several species, while other birds incubate continuously at the expense of feeding. If adverse weather prevails during the incubation period of arctic geese, the female remains sitting. In situations of high wind chill, a lightweight bird loses heat and may even die while incubating.[128]

Some birds, notably grebes and a few wildfowl, cover their eggs while away from the nest. Black-necked Grebes use nest material, which is a ready absorber of solar radiation and keeps the eggs warm on sunny days as well as protecting them from predators. Further heat is supplied by the putrefaction of the underlying nest material, thereby reducing heat loss between incubation stints. As in other grebes, the nest temperature is therefore independent of the temperature of the surrounding water. Many wildfowl cover their eggs with down and moss when leaving the nest. While this protects and hides the eggs and slows heat loss, the eggs may be exposed if strong winds remove the covering.

At high latitudes and altitudes the eggs of certain species are adapted to withstand low temperatures. The resistance of crossbills' eggs is quite remarkable, and other species which either nest in cold climates or breed very early in the year also lay cold-resistant eggs. One or two seabirds and the Swift lay eggs which can withstand a certain amount of cooling to cope with the shorter incubation stints brought about by food scarcity.

Changeover periods at a nest in very low temperatures must be rapid so that the eggs are exposed as little as possible. Nutcrackers do not incubate until the clutch is complete, merely covering the eggs – a strategy similar to that of the Siberian Jay, which may nest in temperatures down to −30°C (although in the latter it is only the female that incubates). The eggs of the Gyrfalcon may be uncovered for only a few minutes on changeover, although at one nest with an air temperature of −35°C the eggs were exposed for a mere 20–45 seconds. Clearly there is a limit to the duration of the interruption which will not harm the eggs, depending upon the resistance to chilling, the nest site and the foraging habits of the birds.

Naturally, protection of eggs and young from high temperatures is found most often in hot climates, but may also be necessary in higher latitudes when summer temperatures reach a hazardous level. Ground temperatures in such climates, especially of sand, reach phenomenal values. The mean ground temperature in Iraq, for example, has been given as 82°C from May to September, occasionally rising to 92°C. This is rather exceptional, however, since the superheated layer is not only reflective, but also subject to convective overturning. It is thus continually being replaced by slightly cooler air from above. Geiger[54] quotes a figure of 60–70°C as being the maximum reached in the Sahara desert and the Middle East – figures that can also be experienced in northern and central Europe and in many areas in North America.

A number of species which breed in open country, notably plovers, bustards and coursers, crouch over their eggs during the hottest parts of the subtropical day, shading them from the sun but at the same time allowing any draught to cool them as much as possible. In temperate climates, a bird may merely incubate for a longer period when the sun is most intense. In all cases it is necessary to maintain the temperature of the eggs at an optimal level. A Common Nighthawk, nesting on a very hot rooftop in the USA at a temperature of 61°C, was able to reduce the egg temperature by 15°C by shading them. The bird orientated her body along the axis of the sun's rays to afford maximum protection to the eggs. The head always pointed away from the sun, but no orientation was shown on cloudy days. Birds whose eggs are not directly exposed to the sun may merely sit by (rather than over) their eggs in very hot weather.

Several species, again chiefly those in low latitudes but also recorded in temperate climates, are known to wet the feathers of the underparts before changing over at the nest in order to effect some control over temperature or humidity or both. Cooling of the eggs by evaporation can be very marked in hot dry air as long as the shells are kept moist; 25°C of cooling can frequently be achieved.

CARE AND DEVELOPMENT OF YOUNG

Nidicolous young, protected within the nest, are vulnerable while uncovered since they are helpless and cannot take cover of their own volition. Nidifugous young can avoid adverse conditions to a certain extent but, until they can fly, this facility is of limited value and their early plumage is of little help. Adult birds must therefore afford the necessary protection, though in many species only one parent is concerned in rearing the young.

The amount of brooding needed by nidicolous young depends upon several factors. Among these, one of the most important is the rate of heat loss of a nestling. Heat loss depends upon exposure, air temperature, and the degree of insulation afforded by the plumage (such as it is), the nest and the remainder of the brood, if any. The insulation of the nest is better for hole-nesting species. Young birds surrounded by wood will be more insulated from the environment than in, say, the crevice of a stone wall, due to the difference in the conductivity of the substance surrounding the nest. The insulation provided by a large brood will also mean less heat loss, so that less food is required per nestling.

Initially, a nestling is very vulnerable to chilling when brooding is interrupted – it becomes torpid, fails to gape, and starves. On hatching, a nestling is cold-blooded (poikilothermic), but slowly becomes warm-blooded (homoiothermic) during the prefledging period, so that its resistance to cold increases with time. Young songbirds are poikilothermic for at least a week, and even in warm weather will die if not brooded at night. The weather is thus particularly important during the days immediately after hatching.

The young of passerines that breed very early can, like the eggs, tolerate quite extreme conditions. Crossbill nestlings have a slow and variable development. They can survive temperatures of −35°C, and the temperature under a brooding hen may be as much as 56°C higher than the outside temperature. The young become torpid when adults are foraging, but recover quickly when brooding is resumed. They are probably better adapted to withstand severe cold than continuous precipitation. The first feathers to grow on young crossbills appear on the exposed upperside of the head, nape and upperparts, and the birds are fully feathered within 11 to 12 days. Other species at the same latitude which breed later, grow feathers either simultaneously over both under- and upper parts, or first on the underparts.

Of nidifugous young, Capercaillie chicks are known to have an especially poor temperature regulating mechanism, requiring brooding for at least twenty days. In persistent cold wet weather, they tend to seek shelter beneath the hen. This often results in starvation, while those that leave to forage become chilled and die of exposure. High autumn populations in Norway occur after dry warm midsummers following an early thaw in April – relating to available food for the hen and optimum conditions for chick survival. In contrast, a similar increase in brooding in other game birds, coupled with a decrease in foraging, does not affect survival unduly. Ptarmigan and Willow Grouse chicks may need only a few minutes per hour foraging to satisfy them, but in low temperature, wind and freezing precipitation, even this figure may not be attainable. The stimulus for Willow Grouse chicks to return for brooding is thought to be the decline in body temperature rather than calls uttered by the hen.

The temperature regulation of nidifugous young generally improves with age, and even when very young, resistance to cold is greater in high latitude species. However, sudden temperature changes and precipitation in cold climates can bring about high mortality, especially as insect food diminishes during such spells. Predation becomes significant and young birds periodically seek shelter beneath the female in cold wet weather.

The waterproofing of waterfowl plumage, in which a layer of air exists between the skin and the water, ensures that young can survive in an aquatic habitat. Wet ducklings and chicks have a lower body temperature than dry birds, and this temperature depends upon the efficiency of the waterproofing and the area of skin wetted.[126] Poorly waterproofed birds lose heat rapidly. Ducklings may seek shelter from rain, and heat is also generated by increased activity. Any factor that disarranges the plumage or lowers its surface tension destroys the waterproofing, and this is most likely in downy young. Much of the mortality of young waterfowl is therefore due to chilling in the first two to three weeks after hatching, and wet, snowy or cold weather in polar regions causes heavy losses in highly synchronised goose colonies if it occurs just after hatching.

Unpredictable food supplies have resulted in two adaptations in nestlings: a resistance to chilling and starvation, through the deposit of fat reserves as insulation and as a buffer against future food shortages; and a reduction in the rate of development, leading to a longer fledging period. These are found in the Swift, some other insectivores and one or two seabirds (see Chapter

12).[87,88,127] The greater the potential effect of a food shortage, the larger the store of fat. Weight variations and growth patterns are probably closely linked to environmental factors.[127] Fledglings benefit from a store of fat both in poor weather and before they become proficient in foraging.

A nestling Swift gains weight until its feathers grow in the fourth week, but a rapid fall in weight occurs when the weather is poor, tempered by torpor – a reduction in body temperature and in the metabolic and breathing rates. Starvation can be tolerated even from hatching, when there has been some nourishment from the yolk sac. One bird with a weight of 57 g survived 21 days with no food, by which time it weighed 21 g. The rate of growth varies with the amount of food received, and the fledging period may show variations of up to three weeks. Weight gains in good feeding conditions can be considerable[88] (Fig. 20).

During precipitation and hot sunshine all young birds are shielded, even

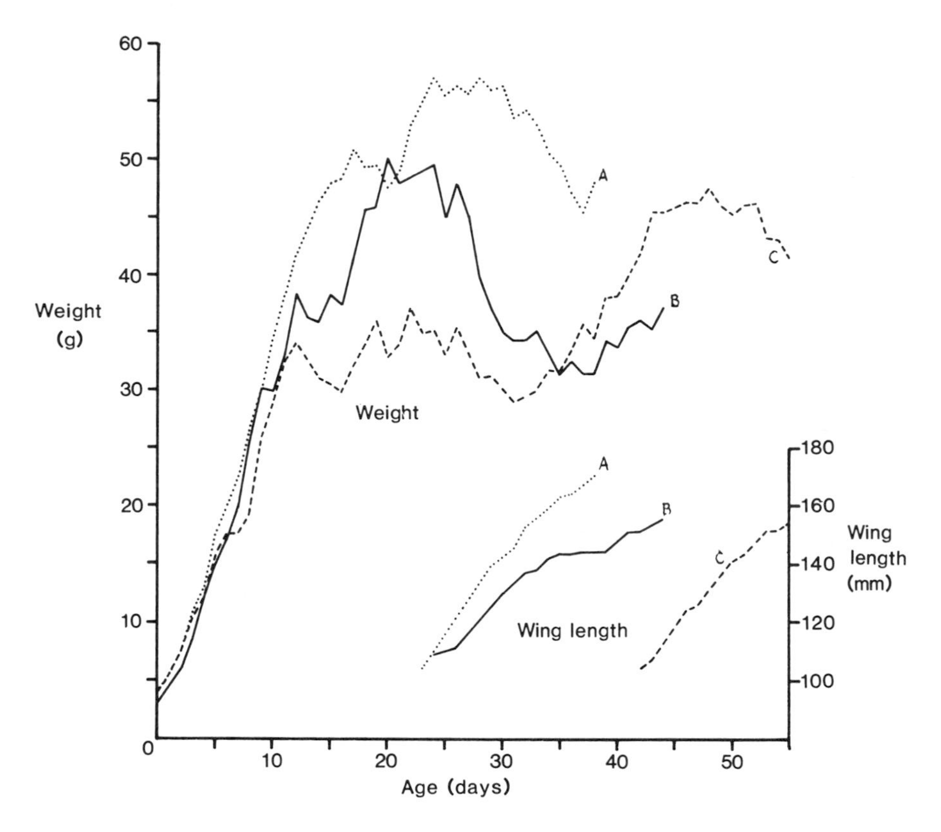

Fig. 20 Increase in weight and wing-length of three nestling Swifts – raised A in good weather, B in good weather at start, bad weather later, and C in bad weather from second to fifth week, then good weather (from Lack 1973)[88].

after regular brooding has ceased. However, as they grow it becomes progressively more difficult for an adult to afford complete protection. Large birds such as raptors, herons and their allies stand over the young with wings spread or drooped (mantling) and tail often spread. Large raptor nestlings may die of chilling in wet weather if both adults need to hunt. In heat, exhaustion may take place if the nest is in a position where no airflow is possible, even when parent birds are attentive. Adults may nevertheless carry water to their young on rare occasions, although this is a behaviour adopted in the extreme by sandgrouse. In some species the belly feathers are specially adapted to carry water over long distances. Up to 40 ml can be held by the male, and although reduced by evaporation up to 18 ml may still be present after a 35 km flight in the cooler, damper, periods of the day.

SUMMARY

In this chapter, I have tried to show how the whole reproductive cycle is influenced by weather.

In late winter and spring, breeding territories are re-established, but often abandoned if cold weather increases maintenance needs and thus the time needed for foraging. Both song and display are also inhibited by poor weather conditions. Temperatures in winter and spring have a marked effect on laying dates, through their influence on the environment and thus the body condition of the individual. Finding sufficient food is a problem for migrants newly arrived on breeding grounds at high latitudes and altitudes, where late snows and frosts delay breeding – often so much so that success is minimal. Flexibility in breeding behaviour is necessarily high in such environments. Although most species lay at a certain time to ensure the maximum abundance of food for newly hatched young, some – particularly those with an unpredictable food supply – tend to lay as soon as there is sufficient food for the female to form eggs. It has been demonstrated that several species begin laying a few days after marked rises in spring temperatures.

Cases of very early breeding in Britain take place in abnormally mild winters when the North Atlantic atmospheric circulation is unusually vigorous. Wetland species show changes in timing and nest site related to water levels, particularly those breeding in drier climates.

The nest and its contents, or the vegetation in which they are situated, are prone to damage or destruction by wind, rain, snow or floods. Clutch size and breeding success are in many cases influenced by the weather operating through the food supply and its degree of unpredictability. This is shown in the extreme by some aerial insectivores, whose food supply varies according to meteorological factors. Their young show variable growth rates; there is fat deposition as an insurance against possible future food scarcity; and an abnormal degree of early thermal independence is developed.

Weather affects incubation and brooding or caring for the young. Protection is afforded by the adults, particularly in exposed situations, and in some extreme cases of high temperature, strategies are employed to enhance cooling of eggs and young. Much mortality of young is due either to starvation, or to exposure to meteorological extremes. Cold wet weather accelerates chilling, and is an important cause of mortality in nidifugous young.

CHAPTER SIX

Comfort

When a bird is not occupied in breeding or feeding activities, it has considerable time in which to engage itself in other, often essential, occupations. These might include feather maintenance or other comfort movements; or environmental conditions might force it to seek shelter; or it may merely do little except rest. During these relatively inactive periods, the bird is still under the influence of the weather – a direct influence rather than through food. During resting periods a bird may sleep, though prolonged unbroken sleep such as higher animals know it, is probably unusual. For birds only able to feed by day, the hours of darkness are necessarily taken up with roosting. Species unable to feed at night include nearly all passerines, and diurnal non-passerines whose feeding is not regulated by a tidal rhythm. A bird's inactivity at night certainly conserves energy, but also means that no heat is generated by muscular activity. It must therefore have enough stored energy to draw upon to maintain its body temperature until it can feed once again; and it is imperative that it finds a relatively sheltered site, safe from predators and to some degree protected from the weather.

Several methods are used to combat heat loss. These include shivering, improving insulation by partially erecting the feathers, covering the legs, tucking the bill into the plumage to breathe pre-warmed air, huddling in tight flocks or deliberately reducing the metabolic rate.[22] When very cold, heat lost to the environment may just be balanced by metabolic heat production, and the lower limit of thermoneutrality is then approached, i.e. the temperature below which internal heat production cannot compensate for heat loss. On these occasions, even a few degrees of amelioration from the environment is critical. In man, the combined effect of wind, temperature and humidity can lead to very serious heat regulation problems. At a given temperature, the wind chill factor increases as the wind speed rises and humidity falls. There is a greater heat loss with a temperature of 0°C and a wind of 12 m/s, than in a calm with a temperature of −10°C. A low humidity leads to additional cooling by evaporation of moisture through respiration and from the skin. This is intensified by wind. Birds are, of course, insulated by plumage, but only to a degree, and the bare parts are mostly unprotected.

ROOSTING

Timing and Technique

The time at which birds go to roost and depart the following morning depends on several factors. The most important of these is light intensity, which in turn is partly influenced by cloud cover. The diurnal behaviour of an individual bird is controlled by its internal circadian rhythm, so that for a day-active species the individual's activity begins later and ends earlier on dark days. The reverse is true for nocturnal species. Roosting birds are very sensitive to day to day variations in light intensity, including those due to snow cover. On the other hand, the necessity for continuous feeding in cold or otherwise adverse weather may delay roosting, and both these factors cause considerable variations in roost timing. The later a small bird leaves its roost the more vital that feeding becomes and the lower the light intensity that will stimulate its departure.

The onset of fog late in the day – a common enough occurrence on a winter evening – brings Starlings to roost as much as two hours earlier than normal. In such poor visibility they may roost closer to the feeding grounds, with fewer birds at the normal roost site. In a very dense January fog in Norfolk, Starlings were unable to locate a roost and flew around in visibilities of 20 m for up to 5 hours in the late evening. Some landed exhausted, but the destination of the others remained unknown. Rooks and Jackdaws fly low when going to roost in fog – presumably to pick out recognisable features on the ground. They roost early in dull overcast weather, spending the last half hour inactive at a pre-roost assembly area, even on days of curtailed feeding. There are records of other birds beating at lighted windows in thick fog – similar to the behaviour shown by migrants at lighthouses in very misty weather (see Chapter 8).

R. A. O. Hickling[72] observed that the gathering of gulls at inland winter roosts lasted some two to three hours before dusk, with arrivals continuing after dark in good weather. In fog, snow or heavy rain the gathering was spread over a shorter period concentrated in the early afternoon. They flew in to the roosts (on reservoirs) at low level in strong winds or when they were to land on ice, but at other times parties spiralled in from a great height. Massed pre-roost flights occurred chiefly in clear skies and fresh winds. Fog forming in late afternoon may prevent gulls finding their roost and, like Starlings, they create makeshift roosts elsewhere. Morning fog inhibits them reaching the normal feeding grounds.

Breeding Swifts roost early when food is scarce in cold weather. When there are young in the nest, however, roosting times are relatively independent of weather and bear a definite relationship to sunset, since the adults need to forage as late in the day as possible.

Crag Martins wintering in Gibraltar show considerable variation in roost timing, according to light intensity and rain.[45] In fine or fair weather, with light to moderate winds, flight from distant Spanish feeding grounds to the cave roosts of Gibraltar begins 30 to 60 minutes before sunset, somewhat later on the shortest December days. In dull cloudy conditions, when light available for feeding is substantially reduced, the birds go to roost up to two to three hours before sunset, and the roosting flight may be prolonged and broken. In extremely poor weather many birds either remain at the roost all day or return very early. This was noted on one November day when heavy rain accompanied easterly winds of 20 to 25 m/s. On a midwinter day, most of the incoming Crag Martins appeared ahead of a marked cold front with heavy rain, which approached from the direction of their feeding grounds; early returns ahead of such fronts were noted on several occasions. The morning departure is delayed in cold weather, doubtless to allow the emergence of insects over the sun-warmed Spanish feeding sites. The variations in roosting times due to poor weather are thus thought to result from food scarcity, reduced foraging efficiency, the necessity to conserve energy and the urge to shelter from rain.

The dependence on foraging is also illustrated by the behaviour of species which use thermals for hunting or travelling to feeding areas. Departures and arrivals of White Pelicans from roosts away from feeding sites in Africa are timed to make use of thermals. Roosts are occupied one to two hours before dark, and evacuated one to three hours after daylight; should poor weather intervene when travelling between feeding grounds, temporary roosts may be formed. Many raptors naturally show a similar timing, although those which use thermals less frequently (or use orographic upcurrents) will leave their roost site earlier and return later. In rain, mist or low temperatures some, such as Red Kites, Golden Eagles and Griffon Vultures, may remain at the roost all day.

Some birds roost singly, others in groups, and yet others huddle closely together in clumps. Communal roosting means larger and fewer roosts. Groups which are not huddling may raise the ambient temperature slightly, but the creation of warmth is probably only a subsidiary function since grouping is found also in warm climates. Such communal gatherings are thought to decrease the risk of predation and possibly convey to a large number of birds information regarding food and feeding sites. Communal roosting also occurs in normally non-gregarious species such as the Blackbird, Wren and Robin (frequently in dense vegetation) and is shown by some raptors, particularly kites and harriers.

Species which roost by huddling together create a definite microclimate in which body heat loss is reduced markedly. Their effective exposed surface area is minimised, and such roosts are formed primarily by small birds which are, in any case, subject to a relatively higher heat loss in cold weather.

Perhaps the most well-known and documented of species which form compact communal roosts is the Wren, although island races do not apparently show this behaviour. In cold weather this tiny bird uses small cavities (such as nest boxes and squirrel dreys) which provide warmth and insulation from low air temperatures. Lined cavities are especially important on nights with critically low temperatures. Wrens usually roost singly until the onset of cold weather and may not survive if, by then, they have not found a communal roost. Low temperature is crucial in initiating and maintaining social roosting, although there are other factors involved.[5] High humidity may also be significant in its capacity to reduce cooling by evaporation. During glazed ice conditions in England in February, 1976, Wren mortality was high; this was attributed not only to food scarcity but also to the inability of the weakened birds to fly up to communal roosts. The numbers of Wrens using one roost can be impressive, the record being 96 roosting together in the loft of a house during a six week period from January to March 1979. The numbers fluctuated according to temperature, and when the peak numbers arrived there was a queue to enter the hole[68] (Table 3).

In winter, Hen Harriers roost communally on the ground. In southwest Scotland, Donald Watson[182] found that the highest numbers occurred in mild, windy weather, while the lowest numbers were on calm, often cold, nights. He suggested that calm frosty weather induced the birds to roost nearer the hunting grounds to conserve energy, and that fine nights, especially when mild in early spring, may possibly stimulate some birds to roost in the breeding territories. He also thought that on calm cold nights, some may have roosted in nearby forests to take advantage of the higher overnight temperatures. The increased numbers in mild midwinter periods were thought to have been compounded by an influx of birds into the area. The harriers' behaviour on arrival at the roost varied with the weather. In strong winds, particularly if accompanied by heavy rain, the birds dropped into cover almost immediately, though in any weather they might rise again to settle in a different spot. They arrived flying into wind just above the vegetation – often close behind each other. On calm nights, the birds frequently glided in slowly from various directions and from a considerable height. On fine windy nights several birds might soar above the roost and, if mild, pairs showed behaviour similar to that on breeding grounds. Like many other communal roosting species, Carrion Crows roost increasingly within their breeding territories from January onwards, totally so by mid March, but exceptionally cold weather may delay these changes by up to three weeks.

Roost site

The degree of shelter offered by a site depends on a combination of factors – the habitat itself, the weather at the time of roosting, and the level to which the overnight temperature falls. The nocturnal minimum air temperature is determined chiefly by air mass type, wind strength and cloud cover and, given relatively constant overall conditions, occurs just after dawn. On a long winter night the duration and magnitude of the temperature fall is greater than in summer. As we saw in Chapter 3, minima also vary according to the nature of the ground surface and, during a nocturnal radiation situation, are higher away from the ground. Roost sites are therefore chosen to provide a favourable microclimate where there is some degree of insulation from the environment, as in cavities or dense vegetation. The choice of site may be critical in winter, and is made according to its shelter value and perhaps its proximity to food sources. Sheltered sites are not so critical for large birds, which lose less heat per unit surface area than small birds.

Cavity roosts afford a high degree of shelter. Several passerines roost communally in sea caves, and Jackdaws and Starlings take advantage of such shelter on exposed Scottish coasts. The Crag Martins of Gibraltar are able to avoid most of the adverse weather in winter by using deep sea caves; only on rare occasions does the wind blow directly into the roosting caves. In one such event, violent gusty winds blew torrential rain into the cave mouths, prohibiting birds from entering freely, and sweeping roosting individuals from perches onto the ground and into the sea. As we have already seen, birds spend longer at the roost in poor weather, and there is evidence from tail feather measurements that abrasion of the tail, from the rough surfaces of the roost, is at a maximum in wet and windy winters.

Overnight weight loss is very dependent upon wind chill, with the shelter value of any site increasing with strengthening wind and falling temperature. Wind speed is probably the more important element. For example, a 73% reduction of wind speed within a woodland Blackbird roost from that prevailing outside the wood was reflected in the weight loss of birds using sites of varying exposure. With a temperature near 0°C, the difference in overnight weight loss between sheltered and exposed groups of birds was 60% in windy conditions, but only 15% in a calm.[156] Weight loss may increase in low humidity when respiratory evaporation is greater, and humidity may even become more important than wind chill.

Water birds seek shelter from rough water by moving closer to a lee shore, and night to night variations in roost position are related to wind speed and direction. Some wildfowl roost ashore in rough weather. Under freezing conditions, Little Grebes will occasionally roost in the air spaces beneath the ice edge of a frozen shore. Night rafts of roosting Goldeneye numbering up to

600 have been recorded in the USA where power station effluent has prevented the freezing of the water. Gulls roosting on reservoirs tend to use the open water, even in gales when ducks have taken shelter. Hickling[72] noted that in a drought they perched on temporary islands, and when the reservoir was partially frozen some birds roosted on the ice edge and others on adjacent water. At all times, safety from predators was paramount.

Just as some species are opportunistic in their feeding behaviour during cold weather, so they are in roosting behaviour. Pied Wagtails are well-known for their selection of warm sites in severe weather, particularly greenhouses. Scottish-ringed birds have been caught in industrial buildings in central England in winter, and even Swallows have survived in heated workshops well into December. Town roosts of Starlings and Jackdaws benefit from the heat island effect of large conurbations, which may be some 2 to 3°C warmer than the surrounding countryside on a very cold night. A disadvantage of town roosting is that the longer flight to feeding grounds beyond the built-up area may nullify any saving in energy, but one of the main benefits is its protection from predators.[66] Nevertheless, the supply of heat from buildings in winter may occasionally raise the temperature contrast between a town centre and the surrounding country to as much as 10°C. The difference in temperature is greatest on a clear calm night, and the energy saving in such circumstances must be considerable.

Flight lines of Starlings to rural roosts change after widespread snow, since feeding is then concentrated within built-up areas, and country feeding is abandoned. This results in marked fluctuations of populations at individual roosts. There is evidence that, in social roosts, dominance works to the benefit of part of the flock. Although the thermal stress is greater near the tops of trees where dominant Rooks roost, the birds are larger and more able to withstand the cold. In strong winds they move lower, or to the leeward side of the roost wood, displacing younger birds to less favourable sites. Thus only in very poor weather is there mortality, and then mainly of younger individuals. Higher roosting birds in social roosts also avoid droppings from above and the possibility of impaired plumage quality, which may be fatal in heavy rain.

Species resident in cold climates adapt to their environment by burrowing into snow to seek a favourable microclimate. Many gallinaceous birds and some passerines use snow holes for shelter and insulation, even in Britain. With little snow, Ptarmigan will concentrate on the available patches and in cold, snowy springs, moult may be delayed in order to retain the cryptic, and perhaps insulative, properties of the white winter plumage. They burrow deeply only in calm weather, since deep burrows would fill with drifted snow very rapidly in a wind, and in burrows they also avoid the extremely low nocturnal temperatures that are reached on and just below the snow surface on a calm night. In windy situations, shallow holes in exposed places are favoured, minimising the risk of being buried, and individuals move during a blizzard to avoid this possibility. The depth of snow holes in Scotland also varies with the snow structure, and Ptarmigan appear to burrow to as high a temperature as they can, bearing in mind the constraints due to wind. In hard snow they use hollows only two to three centimetres deep, or shelter behind stones. In slightly frozen snow, or granular snow (snow pellets), the hollows are 5 to 7 cm deep and occasionally up to 20 cm; in powder snow they may be as deep as 30 cm.[180]

Red Grouse make hollows rather than burrows, and face the wind; after cold nights their backs may be covered in hoar frost. To conserve energy they and other related species may remain all day in sheltered sites during extreme cold. Pheasants roost on the ground and in trees (there are records of up to 15 on one branch in poor weather) and in extremely cold conditions one bird remained on its perch continuously for 42 hours. The importance of snow holes for roosting is illustrated by grouse in Alaska which could not burrow through a covering of glazed ice, suffering high mortality as a result.[103] On the other hand, Prairie Chickens in USA died while roosting beneath the snow, after being imprisoned by glazed ice forming on the surface. In climates with very cold winters, the distribution of Capercaillie, and perhaps of other grouse, is limited to areas where snow is deep enough to burrow down from the low surface temperatures. Temperatures may be as low as −50°C at the snow surface, but only −5°C 60 cm below the surface. Arctic passerines such as Redpolls and Snow Buntings also burrow to roost in warmer conditions. In Siberia, Willow Tits may roost in snow tunnels, dug either by themselves or rodents, and after heavy snowfall they may also need to dig themselves out.

RESPONSE TO LOW TEMPERATURE

In addition to the search for a favourable microclimate in which to roost, a further adaptation shown by a few species in response to cold weather is that of a reduction in body temperature (hypothermia), which minimises the amount of energy used during roosting. I discussed this in respect of some young birds in Chapter 5, in which torpor occurs. Adult birds of certain families also have this ability to become torpid. There is a distinction, however, between hypothermia as an involuntary reduction in body temperature arising from prolonged cold and a depletion of fat reserves due to food scarcity (a prelude to death, in effect) and that which is an adaptation to minimise the effects of such events (see Table 4).

The former is a response shown by all species close to death by starvation in cold weather. As a bird's internal temperature falls, so does its metabolic rate. Hirundines are particularly prone to hypothermia. They lose heat easily and may huddle during a cold day (Chapter 4) with an accompanying drop in metabolism. In these cases they cannot feed and, if the situation does not improve, death ensues within three or four days. Huddling has been recorded in bee-eaters and swifts. Fifty to sixty Swifts were seen clinging to each other inside the wall of a tower during a thunderstorm, and even greater numbers (up to 200) have been recorded. In these clumps each individual seeks to avoid the outside positions, resulting in the unsuccessful birds succumbing to the cold.[88]

Hypothermia as an adaptation for survival is found in some nightjars, hummingbirds and swifts, and occasionally small resident birds in cold climates. In this method of energy conservation, the body temperature is depressed but regulated at some lower level. A more extreme state is torpor, in which the heart, respiratory and metabolic rates are greatly depressed, culminating in the hibernation practised by one or two species. Through experiments with small northern passerines (Great Tit, Greenfinch, Brambling, House and Tree Sparrows and Redpoll), Steen[165] found that the birds became acclimatised to cold. However, their reactions suggested that extremely low overnight temperatures are avoided by seeking shelter, and that regulated hypothermia is a second defence against abnormally cold weather.

I mentioned earlier the deliberate covering of the legs, feet and bill to reduce heat loss from these bare extremities. In arctic conditions, the legs and feet are exposed to low temperatures for prolonged periods. Frequently they are wet from snow or water, thus lowering their temperature even further in flight with a risk of icing. Some birds have feathered legs and feet, but the chief mechanism by which the tissue temperature is maintained above 0°C is a counter-current heat exchange in which the blood flow to the legs and feet is periodically increased. One of the more frequent leg positions of resting birds in low temperatures is that in which one leg is withdrawn into the plumage, alternating from time to time with the other leg. This position is normally shown by shorebirds, but occasionally by passerines. Passerines and wildfowl more commonly sit on the ground or perch and cover both legs with the plumage. In temperatures of −40°C, Snow Buntings will forage with flank and belly feathers lowered to cover the legs.[115] In cold weather wetland birds have been observed to draw their legs and feet forward in flight. Coot, Redshank, Spotted Redshank, Dunlin and Black-headed Gull are known to use this position as soon as they take flight – presenting a shortened appearance in those species which normally trail their legs.

RESPONSE TO SUNSHINE

Sunshine elicits two different reactions: shelter if extreme, or comfort movements when warmth is required. Unusually high temperatures in intense sunshine provoke various responses aimed at reducing the body temperature. The plumage may be sleeked to reduce insulation or ruffled to catch any breeze, and a metabolic response may be invoked in which the body temperature is raised (hyperthermia) in an attempt to maintain heat loss to the air. However, if the air temperature is too high, death finally ensues from heat exhaustion (see Table 4).[211]

Many birds of open country do not actively seek shelter from the direct rays of the sun. M. D. England[47] watched a Great Bustard sitting, in great heat, with its back to the breeze so that cooling extended to the skin under the erected feathers. Temperatures are highest on the surface of the ground and, as already mentioned, some surfaces (particularly sand) may exceed 60°C. Even stones and rocks raised a little way off the ground are at much lower temperatures, and ground feeders in arid country, such as wheatears, will perch on termite heaps or the low branches of acacia trees to avoid extreme heat. Palearctic migrants wintering in Africa regularly experience air temperatures of 30°C or more and will rest in the shade for two to three hours, sometimes up to six hours, in the middle of the day.[107]

Body positions in high temperatures may be modified so that the skin of the upper and under surfaces of the wing lose heat by conduction and radiation, and, additionally, the blood circulation to the legs may be increased to enhance heat loss. Some species achieve heat loss by sitting on the tarsi and spreading the wings slightly, or raising the wings to expose the flanks. A common method of heat dissipation is panting, undertaken when the body temperature reaches a certain threshold. It is often accompanied by an increased rate of breathing – the first stages of heat exhaustion if the resulting heat loss is inadequate. Panting allows cooling by respiratory evaporation, which is essential to an animal that cannot sweat. Some larger birds, notably gannets, cormorants, herons and their allies, enhance evaporation by fluttering the gular pouch. The Gannet also tilts its head to expose the naked black skin of the throat to allow heat loss (black surfaces both absorb and radiate heat efficiently) and exposes the webs of its feet. A number of species excrete onto the webs to augment evaporative cooling.[114] The feet of others are of great importance in the loss of heat, both by cooling in flight and in water. The Mallard and Herring Gull, among others, are able to lose a high proportion of their heat production by adjustments to the blood circulation in the feet.

Although the sun may be avoided if possible when its heat becomes uncomfortable, it is sought if warmth is needed, and a wide variety of species sunbathe. Sunbathing may consist merely of seeking a sheltered spot in the sun after a cold night, or of a high intensity behaviour in which characteristic postures are adopted. The most frequent postures are those in which the wings and tail are spread, feathers ruffled and the maximum surface area exposed to the sun. Sunbathing may be carried out on the ground or on perches in trees or on buildings, but usually in a place safe from predators – in which case the behaviour may last for a considerable time. Some species raise one wing so that the sun strikes the underside and the flanks. Explanations of the motives for sunbathing have been rather controversial. Suggestions include the sunlight forcing ectoparasites to move to accessible areas where they can be removed; the manufacture of vitamin D in the skin; and, most likely of all, a simple temperature response to warm the body.[79]

D. C. Houston[74] offered a further hypothesis, particularly for large static-

soaring birds. He had observed vultures *Gyps* sunning for periods of a few minutes upon landing, and postulated that the temporary deformation of flight feathers during prolonged soaring would be detrimental to flight. Since such birds require a highly efficient aerodynamic wing, especially during the difficult post-feeding take-off, he suggested that sunning would rapidly restore the feathers to the optimum condition, and offered some experimental proof.

RESPONSE TO PRECIPITATION AND WIND

The presence of rain or snow occasionally elicits bathing and many species bathe in snow, or in grass or foliage made wet by rain, drizzle or dew. Snow bathing – deliberate movements in which the head is buried and the snow, usually when powdery, is flung over the back – appears often to be additional to normal bathing, rather than a substitute on occasions when standing water is unavailable. It has been observed even when unfrozen standing water is present nearby. Grass may also be made wet through deposition of fog or mist droplets or the process of guttation – commonly confused with dew, but quite distinct. Guttation appears as large droplets of water at the tips of grass leaves, and is a physiological process resulting from root pressure within the plant. On the other hand, dew forms during nocturnal cooling and results in small droplets of water covering leaf surfaces.

Although bathing is a controlled wetting, there are frequently occasions when precipitation causes involuntary drenching of the plumage. Initially, shelter may be sought from both wind and precipitation. In open situations, perching birds minimise the effect of strong winds by sitting or standing with head into wind to avoid heat loss from plumage disarrangement, and to allow the natural stream-lining of the body to reduce friction and buffeting. Water birds tend to seek shelter from rough water and some ducks form compact rafts facing the wind. Ground perching birds shelter behind prominences such as stones, clods of earth and tufts of grass, or in furrows or holes, particularly if the wind is accompanied by precipitation or blowing sand or snow. Prowse[143] watched Skylarks take shelter from a blizzard in the lee of grass tussocks (one bird per tussock) and shuffle the feathers regularly to prevent snow settling on the body; normal feeding behaviour was resumed when the snow ceased.

Most birds do not actively avoid light rain, but seek shelter when rain becomes heavy or driving. Sheltering is widespread among tropical birds which frequently experience intense, but short-lived, downpours.[80] If wetting cannot be avoided, behaviour and postures are adopted to rid the feathers of excess moisture. The plumage of most birds is able to shed water for at least a short while; individuals repeatedly exposed to wet and cold conditions are more able to maintain their body temperature than those not so habituated, probably being adapted physiologically to such weather. Rain drops may be removed by wing flapping and body shaking, often carried out in flight by ducks and gulls. The plumage can be sleeked and a position adopted in which the least surface area is presented to the rain; this is best developed among tropical species unable to take shelter from torrential downpours. Despite the belief that Cormorants and Shags stretch their wings to dry them their plumage is as waterproof as other seabirds, although the wing feathers are more easily wetted. The wetting of raptor plumages seems

to inhibit hunting but, in Africa, Hobbies continue to feed in the rain on emerging termites, shaking themselves dry every few minutes. In southern Spain a flock of about 35 rather bedraggled Griffon Vultures was seen drying in the sun after rain, some with wings outspread like Cormorants. This appears to be related directly to the feather maintenance of soaring birds mentioned in the previous section.

SUMMARY

In this chapter I have described the effects of weather on comfort activities. To a degree, weather determines roost timing and site, the latter selected to give maximum protection from weather except in those species for which avoidance of predation is more important. Favourable microclimates include communal roosts, especially those which involve huddling with other individuals. The direct reactions of birds to temperature, sunshine, wind and precipitation are discussed, and their effect on thermoregulation.

CHAPTER SEVEN

Migration: inception and progress

Migration has been defined as a regular seasonal journey between two areas, performed by a species which thereby reproduces or survives more efficiently than if it stayed in one place all the year. It occurs primarily in response to seasonal changes in food resources, and in most instances has evolved over lengthy time periods. Food abundance varies with climate. In temperate and polar regions the changes are caused by warm and cold seasons, in the tropics by wet and dry seasons. Some fluctuations are more indirect. For example, certain predators, both seabirds and land-based birds of prey, migrate according to the abundance and accessibility of vertebrate prey, some of which move as a result of the climatic changes mentioned above. Within the sphere of migration there is a whole range of movement from long distance trans-continental or oceanic migration to short distance wanderings.

The precise timing, duration and route of migratory movements are partly determined by meteorological factors. These may affect both inception of the movement and the performance of the migrant in flight. In mid and high latitudes, autumn migration frequently begins well before the dwindling of food supplies so that fat reserves for the flight can be accumulated. Certain weather situations stimulate movement in those birds physiologically ready to depart, i.e. with a strong migratory urge. The autumn passage south from northern regions is often leisurely, but in spring there is an urgency to reach the nesting grounds in time to breed successfully – a timing vital to those species breeding in polar regions; at this season birds move much more rapidly. The more rigorous the climate of the breeding grounds, the fewer the species that remain in winter, and for that reason, maritime climates support a greater variety of resident species in winter than do continental climates. Once on migration, the weather can ground a migrant, delay or hasten it, or deflect it from its heading – or even kill it. In this and succeeding chapters, I do not intend to write a treatise on migration, but merely review the influence of weather upon its inception and progress, and the advantages and disadvantages that it bestows upon a migrant. Most of the visible migration studies that have been undertaken have concentrated on the more easily observed autumn migration. At this time, the birds are leaving an area soon to change from one which is hospitable and food-rich, to one in which the food is scarce and the weather periodically poor.

As long ago as the turn of the century, ornithologists observed that migrants appeared at certain places in certain weather conditions. In the late 1940s and early 1950s, observatories set up to study migration began to analyse visible movements in the context of the meteorological conditions pertaining at the time. When radar became sufficiently sophisticated after the Second World War, it was realised that migrants could be watched on radar screens, independent of weather or time of day – opening up new vistas in the study of migration. It became clear that 'normal' migration could be watched, rather than the more atypical movements shown in 'falls' of migrants, which had been to the fore in previous studies. Movements could also be observed at several different altitudes, and at first the results of radar studies conflicted considerably with those of visual studies. Staunchly held theories tended to be dismissed, but it eventually became obvious that the two types of analyses complemented each other, bearing in mind the disadvantages inherent in each. For example, radar 'sees' migration in progress except at low levels below its horizon, but problems arise in identification. Larger species and flocks show as coherent echoes, but there are often difficulties in identifying small migrants. Visual methods can only see low level movements, but have the advantage of being able to identify the species accurately. When the methods are combined, migration shows itself to be truly complex. Some theories remain controversial, but in recent years, the controversy is not so much about the direct influence of the weather upon a migrant, but more about the degree to which it interferes with the migrant's ability to find its way. I shall not discuss navigation at length, but only in so far as it has a bearing on the meteorological aspects of migration.

NAVIGATION

Although different individuals react in many a diverse manner at the onset of or during migration, in general a migrant heading for a definite goal requires conditions in which it can at least set out with the prospect of being able to orient successfully, i.e. to take up and maintain a compass direction. The varying reactions that individuals show are related to age and experience, and not all populations of one species react in the same way. This poses problems for migration watchers and analysts, since one is never sure of the origin of the birds under observation, and several populations may be on the move at any one time. Some species are less affected by the weather than others, and to add to the complexity, there are non-oriented movements which obscure the picture even more.

There has been, and still is, much interest and controversy regarding the methods by which a bird navigates.[199] It is probable that an experienced migrant integrates stimuli from several different sources, each with its own value under particular circumstances. The use of terrestrial magnetism, polarised light, infrasound, olfaction and topographical features have been proposed, with some confirmed experimentally and some in the field. The method which concerns us most of all, in a discussion of the meteorological influences on migration, is that of celestial orientation. Migrants are believed to use the earth's magnetic field or the setting sun, or both, to calibrate their compass sense in order to orientate by the sun and stars. Any way in which these astronomical cues are obscured causes a migrant to delay or rely more heavily on alternative methods.

INCEPTION IN AUTUMN

To be able to set out on a migratory journey with the sun or stars visible requires at least part of the sky clear of thick cloud. There are, of course, a great diversity of weather situations that give such conditions; some are short-lived or affecting a restricted area only, others persistent or covering a wide area. It is probable that birds assess the current weather by observing and sensing the signs just as a human observer does. The combination of several then initiates migration in those birds whose migratory urge is strong. It is the more persistent meteorological situations with suitable conditions that are most advantageous from the bird's point of view. If such optimum conditions last for some time over its route, the migrant has a greater chance of performing the first part of its journey successfully. In autumn, those in higher latitudes are more likely to experience adverse weather. Many migrants travel in flocks, even species normally solitary at other times, and it has been suggested that this behaviour is adaptive and beneficial to long distance migrants, since flocks are able to navigate more accurately than single birds.[171] The mean flock size of several species tends to be larger under overcast than under clear skies, and contact calling becomes more frequent.

The synoptic situations which give the most favourable migration weather are those providing substantial areas of clear skies and favourable winds over

the area of origin of the migrant, particularly if persistent. From the discussion on atmospheric circulation in Chapter 1, it is obvious that subsiding air over a wide area will tend to provide cloudless weather. We have seen that this gently sinking air occurs in anticyclonic situations, and also in the regions of subsidence associated with the vertical circulation of depressions, i.e. in their polar airmasses. Admittedly these polar airstreams over the relatively warm oceans, and also over warm land in early autumn, produce convection, but in Chapter 2 I pointed out that convection cells themselves are features with a vertical circulation in which subsidence occurs between the cells. Thus cloud cover is by no means complete. Two other factors of importance are related to these synoptic situations. The first is temperature. Clear nocturnal skies in a high pressure zone allow radiation loss from the earth, and temperatures fall. In polar airmasses, a fall in temperature is brought about by the advection of cold air over the earth's surface. These falls in temperature may tend to stimulate the migrants to leave their breeding areas, particularly the small insectivores. The second, and probably more important, factor is wind. In anticylonic pressure patterns, winds are generally light and often variable in direction. This enables a migrant to fly with little risk of drift from its preferred heading or track. At the onset of a polar airmass, i.e. after the passage of a cold or occluded front, winds are frequently very much stronger. However, they often have the advantage of blowing from favourable directions. Migrants can then move with a downwind component at a speed greatly exceeding their own still-air speed, so that journeys are hastened.

There are other meteorological elements that stimulate departures, especially of species which do not need to migrate until food becomes unavailable. Wildfowl of cold climates, for example, may remain until forced to move by freezing waters or snowstorms. In a mild winter, movement may be as late as January or February, or even not at all.

The importance of such synoptic patterns varies with the physiological state of the bird. Early in the season, there is little real urgency to migrate, but later – especially if the synoptic situation has been adverse for some time and has delayed the inception – those patterns assume great importance, and huge waves of birds leave within a very short period of time. The greater the delay, the more readily will a bird depart in suboptimal weather. The same applies to onward passage of grounded migrants.

We can now identify particular synoptic situations in which these patterns arise. In Chapter 1, I described the existence of blocking anticyclones – those persistent high pressure areas in mid and high latitudes that block the normal west to east flow of air round the hemisphere. They divert disturbances such as frontal depressions, thereby ensuring that anticyclonic conditions, and all

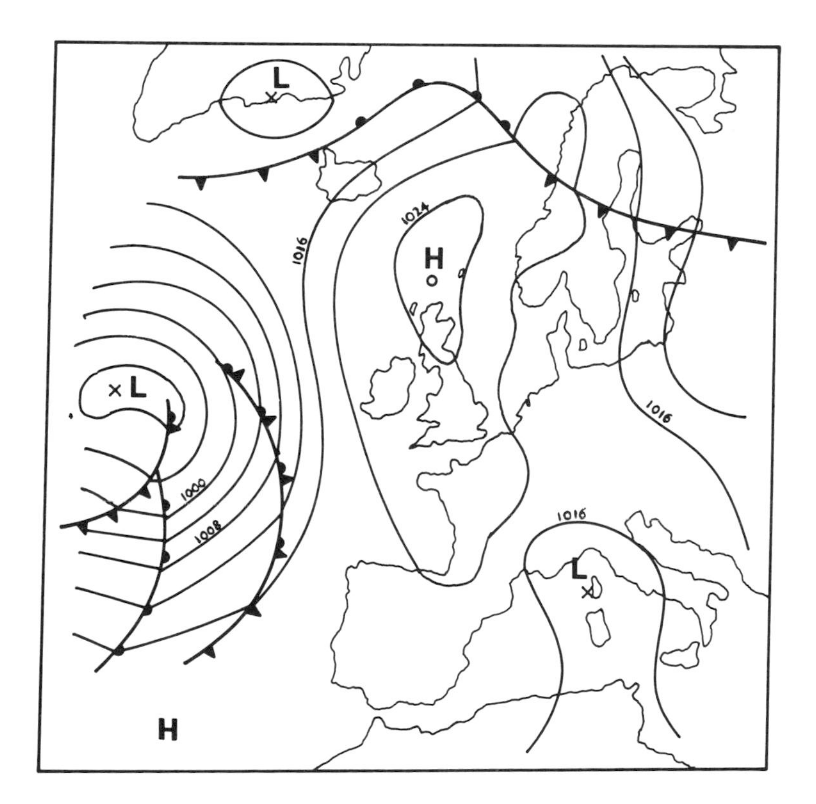

Fig. 21 Blocking anticyclone over Britain, 22nd September, 1977, 1200 hours. Ideal migration weather over Britain and western Europe, although windward North Sea coasts remained cloudy.

they portend for migrants, exist over a particular region for some considerable time. Blocking anticyclones tend to be more frequent at certain times of the year. Their position varies, but if one remembers the details of anticyclonic circulation, it will be clear that the most ideal migration weather will exist within and on the eastern flank of an anticyclone or ridge, i.e. where there are winds with a northerly component. This latter position is often the same region as that affected by cold outbreaks from the north behind cold fronts. Autumn is a particularly favourable period for blocking 'highs' in longitudes eastwards of 20°W (Fig. 21). The frequency increases after August to reach a peak in October and November, although the incidence varies greatly from year to year. These highs can be very persistent, with a mean duration of around 16 days and frequently lasting as long as three weeks, but during their life they may drift, often westwards against the earth's spin, and therefore the type of weather over a particular area will not be as persistent as the high itself.

The other pattern, one of cold polar air moving in a southerly or southeasterly direction, is one which occurs most frequently to the rear of a cold front behind a depression (see Fig. 25). Again, reference to Chapter 1 will show that the amount of cloud depends upon the underlying surface, since convection will arise where warmth is applied to the invariably unstable airmass. Over the sea, which in autumn is usually at its warmest, convective cloud is most abundant, its depth and activity dependent upon its position relative to the depression centre. The most unstable air, and therefore the maximum amounts of deep cloud and the greatest chance of further cloudy disturbances forming in the airstream, is found not close behind the depression (where warm air has been swept in the vortex) but some distance to the rear. Further away again, where subsidence is more pronounced, the depth of convection decreases. Cloud may become more layered and with fewer gaps, though shallower. Over land, too, any convection dies away late in the day so that nocturnal migrants find clearing skies under which to depart. Over the whole of the region away from the depression, winds in the polar air may blow with a northerly component, but if the depression is moving on a track other than approximately west to east, or should it be slow moving with the fronts sweeping to the south of the centre, winds can blow from directions as far back as southwest.

On some occasions, the high pressure zone behind a travelling depression may not be much in evidence ahead of the next low pressure system. Known as a col, it is often merely a restricted area of light winds and broken shallow cloud. For that reason, it has been identified as a pattern associated with migration, though not at all persistent.

Horizontal visibility in polar airmasses is invariably excellent, posing no problems for migrants, but in the light winds or calms of anticyclonic weather, nocturnal radiational cooling over land initiates widespread fog at times. This is usually fairly shallow and is at its most dense around the coldest part of the night near dawn. Hence it is some time after nocturnal migrants have departed (under normal circumstances, departures are densest around sunset and early in the night) and since skies are generally clear above the fog, navigation is unhindered.

Studies of autumn migration relating the inception and progress of passage to weather have been made in many countries, both by means of radar and visual observations. Recent radar studies have been undertaken largely through interest in the hazards that migration poses to aircraft. The chief British contributors to our understanding of weather and migration have been, amongst others, Kenneth Williamson and David Lack. Reference to the bibliography will reveal names of other workers in Europe and North America who have made similarly valuable contributions.

Kenneth Williamson, while director of the bird observatory on Fair Isle in the late 1940s and early 1950s, correlated arrivals of Greenland and Icelandic migrants in northern Scotland with the concurrent weather situation. He hypothesised that[186] the birds followed the windflow in polar airmasses to the rear of large Atlantic depressions, so that they could obtain the maximum benefit from tail winds. He called this 'cyclonic approach'. The inception of the passage is in the clear weather of the post-depression ridge of high pressure, and the persistent winds that they meet further south or southeast give them ample opportunity for fast downwind flight. Greenland birds originate in a region where dense cold air over the ice-cap is conducive to the formation of high pressure which affects the adjacent coasts. September arrivals of these birds on the north and west seaboards of Britain, such as Wheatears and Redpolls of the Greenland forms, and Lapland Buntings, coincide with the northwesterlies to the rear of a depression over or to the east of the Denmark Strait. Williamson attempted to correlate the weights of newly arrived Wheatears with the hypothetical distance flown. He main-

tained that the lightest birds were those which had undertaken the longest journeys of 2,500 km or more, and had lost 35 to 40% of their assumed departure weight, i.e. those using winds in a depression that gave a sharp curvature to their trajectory. He thought that heavier migrants had used a more direct sea crossing with a depression further away to the north-east. A northwest to west airstream to the north of an east Atlantic ridge of high pressure assists Icelandic species such as Redwings and Merlins, in addition to the species mentioned above. Presumed passerines, thought to be Wheatears, Redwings and Meadow Pipits, have been watched on radar approaching the Outer Hebrides from the NW and WNW in all these northwesterly types of situation.[91] Migration may also be initiated in cols, where broken cloud and the absence of wind makes successful passage possible – albeit shorter hops from Greenland and Iceland via the Faeroes and Shetland. Influxes of Lapland Buntings from Greenland have also been correlated with an eastward moving ridge of high pressure – the fine weather of which the buntings were able to follow across the Atlantic to Britain.

Redwings have made very long passages with a deep November depression over Britain. In this instance a slow moving depression maintained persistent strong north to northwest winds over the whole of the eastern North Atlantic, and downwind flight resulted in the recoveries of three dead or dying Icelandic-ringed birds in southwest France and north Spain. Williamson postulated[191] that these migrants have all adapted to make long and fast oceanic flights as part of their normal autumn migration, although such a long passage from Iceland to Spain in one unbroken flight may be unusual. Shortly we shall see how small North American migrants make similar long oceanic migrations.

Winds change little with height in the persistent northwesterlies of a deep occluded Atlantic depression, since by this stage its circulation has spread to high altitudes. Altitude of flight therefore makes little significant difference to the track or speed of a migrant. The upper wind trajectories are also cyclonically curved, but whereas low level winds decrease as the post-depression ridge approaches, the upper winds often strengthen in the jet stream ahead of the next frontal system, where the warm air aloft is carried well ahead of the surface features. This means that migrants departing in the ridge of high pressure can still obtain the advantages of fast downwind flight if they climb to higher altitudes. Small migrants probably remain low, but it has been proven that some larger species – notably wildfowl – do fly at altitudes where they can use strong high level winds. An excellent example is that of a party of swans, probably Whoopers, observed by an aircraft in December 1967 over the Inner Hebrides. The flock was flying south at an altitude of 8,200 m, and was watched by radar to descend towards the north of Ireland. The swans were in northerly winds of 50 m/s at the edge of a jet stream where the high level warm air from a western Atlantic depression was beginning to override the polar air of the previous system. The swans were assumed to have originated in Iceland in a ridge of high pressure at dawn, and

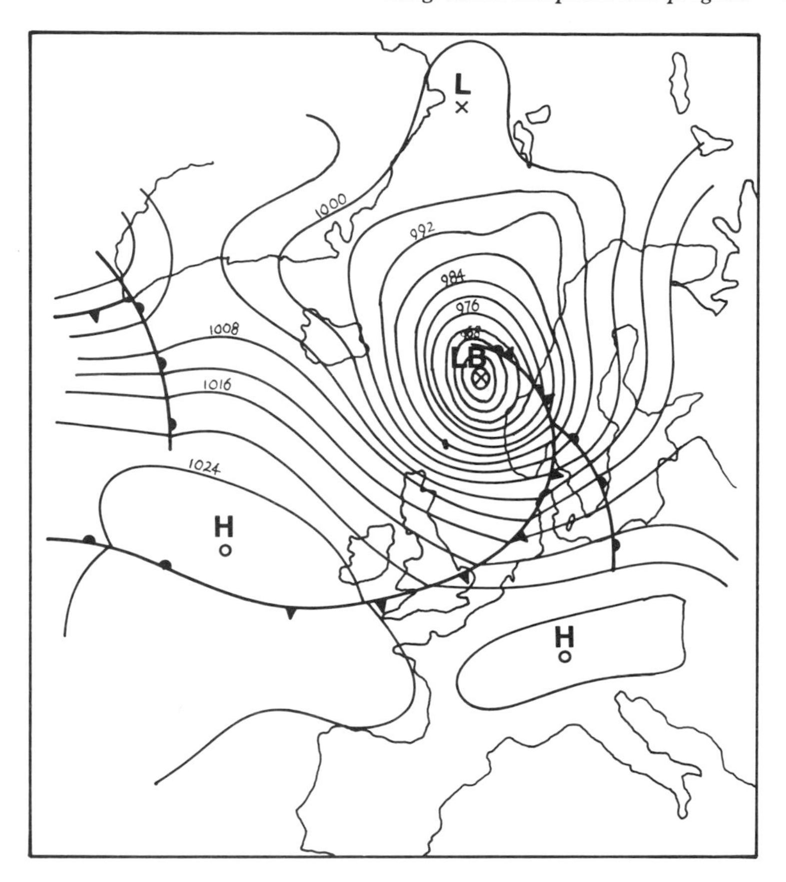

Fig. 22 Cyclonic approach to Britain, 16th September, 1978, 2400 hours. Compare with satellite image in Plate 1. Large numbers of Pink-footed Geese arrived in Scotland on 17th, after departing Iceland in the clearing skies and following winds. The westerly gales associated with Low B brought a large flight of Leach's Petrels into coastal areas of NW England and Denmark.

calculations using upper wind data showed an estimated flight time of seven hours (see Figs. 23 & 24).[43] Despite a temperature of −48°C at that position, the observation suggests that it is clearly feasible for other large birds to use such jets regularly, though definite evidence is lacking. Many geese fly at heights up to 6,000 m, on occasions out of necessity to cross high mountain ranges. Nevertheless, Lesser Snow Geese do so without this necessity over their North American flyway, but they have a preference for lighter winds than are found in jet streams. Several species of geese which migrate to the British Isles from Iceland, Greenland and the Canadian arctic islands, make use of cyclonic approach (Fig. 22). Flocks have been observed in favourable

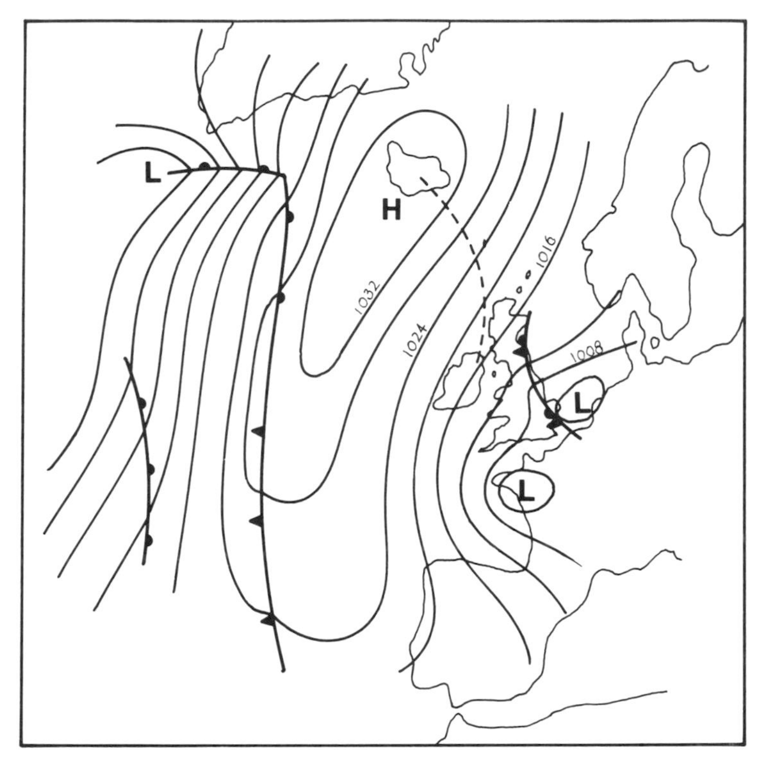

Fig. 23 9th December, 1967, 1200 hours.
- - - - - - Probable track of southward migrating Whooper Swans.

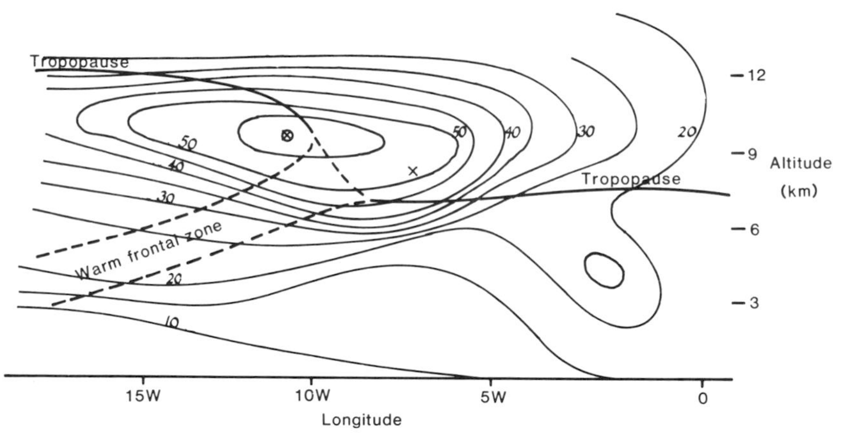

Fig. 24 West-east cross-section of wind speed across northerly jet stream, afternoon of 9th December, 1967 (see Fig. 23).
———— wind speed in metres per second
⊗ core of jet stream
× position of migrating Whooper Swans located at 56½N 7W

weather over northwest Iceland, together with arctic waders, heading towards strong north to northwest airflows over the east Atlantic.[57] West Greenland and Canadian White-fronted and Brent Geese cross the Greenland ice-cap, which rises to 3,500 m.

In northern Europe and eastern North America, post-cold frontal air-masses with winds between northwest and northeast have been identified as being one of the most important situations to initiate passage. In southern Sweden, most late autumn nocturnal passerine migrants leave in mass

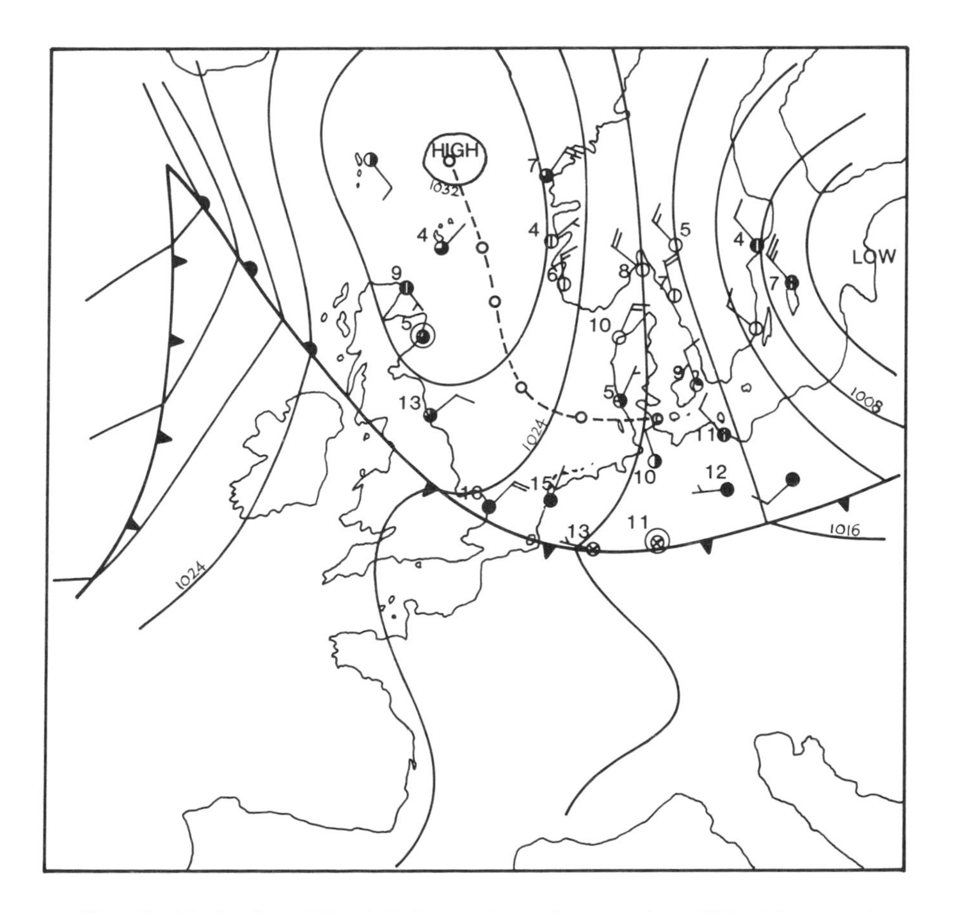

Fig. 25 4th October, 1971, 0600 hours. A massive migration of Woodpigeons and passerines was observed in the northerlies over southern Sweden. A typical transitory situation conducive to mass migration from Scandinavia. The track and position of the anticyclone are shown at six-hourly intervals. The cold front at 1200 hours on 3rd October lay from Stockholm through the Skagerrak to northern Scotland. By 1800 hours on 5th October westerlies prevailed over Sweden – polar anticyclones are frequently short-lived. Temperatures in °C.

movements behind a cold front. The chief autumn movements of the diurnally migrating Woodpigeon are also correlated with the associated strong northerly winds, which hastens their passage to winter quarters. Their dependence upon a tail wind is especially marked when flying over the sea. In Scandinavia, suitable post-frontal weather arises with a high pressure zone over Britain or the North Sea, with a depression having moved east across the region (Fig. 25). Since such ideal situations may not be very frequent, a large proportion of some migrant populations depart at that time.

Passerine migration from eastern Canada and the northeastern USA also takes place chiefly behind cold fronts, and passage oriented southeast to southwest is heaviest in winds with a following component i.e. between west and east through north.[145] A statistical analysis[1] revealed that the wind direction, the 24-hour change in temperature, and an index of the synoptic situation accounted for over 50% of the night-to-night variability. The heaviest passage was associated with northerly winds, falling temperature, and the onset of anticyclonic weather behind a cold front, although the conditions which initiated the passage were not identified. Long oceanic crossings southwards across the western Atlantic to the West Indies are made by some small passerines; these would not be physically possible without wind assistance, since fat levels are not sufficient to allow the distance to be covered under their own power. Fat accumulation by long distance migrants is very closely coordinated with the peak migratory urge. It is greater under clear skies than under overcast, and largest of all when the rate of migration is at its highest.

The existence of tail winds is obviously an important factor in the initiation and maintenance of migration. Radar studies in eastern and southern USA showed that autumn migrant landbirds flew with the wind regardless of direction and speed, but they nonetheless selected winds so that the largest movements occurred in a southward direction. Wildfowl and waders, however, flew in their preferred direction regardless of the wind – showing less selectivity and indicating that they undergo a reduced risk of drift with their stronger flight. Nevertheless, other evidence suggests that wildfowl await favourable tail winds before migrating, tending to avoid crosswinds if at all possible. Pre-migratory flocking has been recorded in ducks in Ohio, in which compact rafts of birds facing into the wind awaited clearing weather. The flocking was considered to enhance mutual stimulation when migration conditions became suitable.

Shelduck departing England on moult migration in late summer leave on fine evenings with little or no cloud, good visibility and a following wind. The wind direction is critical. Poor visibility, adverse weather and headwinds stop all activity, and marked drift occurs if strong crosswinds are encountered. Conflicting observations are not unusual in migration studies, of course, but it is of interest that birds as large as geese and swans are influenced by the wind to this extent, since they are heavy and deflected much less than small birds.[134]

In high latitudes southeast-oriented waders are able to take advantage of wind systems. For example, northeast Canadian and Greenland species, such as Knot and Turnstone, migrate to NW Europe using when possible the same northwest windflows as wildfowl and passerines. Those from Alaska and northwest Canada migrating southeast between July and September may use the strong northwesterly upper winds (at, say, 3 km) which blow at this time from the Beaufort Sea to Ontario. Depressions forming in the lee of the Yukon mountains are steered by this flow along the Arctic front – which divides very cold arctic air from less cold polar air. The strong upper flow meets the main polar front jet stream in the Great Lakes area. This latter jet carries depressions formed to the lee of the Rockies across Labrador where they occlude and turn northeast. It is these depressions that give the northwesterlies so favoured by oversea landbird migrants.

There are a few individuals that move upwind – the proportion of any population being small. Among species that do so regularly are the aerial insectivores. We have already seen (Chapter 4) that Swifts fly into the wind in summer weather movements. Swallows have been observed to migrate upwind in autumn, although this does not seem to apply in spring. They may do so to facilitate feeding at low levels on the wing, since slow flight downwind in the turbulent boundary layer runs the risk of reaching stalling speed in gusts and lulls (see Chapter 2).

The rather lighter winds further behind a depression, particularly those in the clear skies nearer the ridge, are also used by large numbers of migrants. Many SW-oriented birds in western Europe are more closely associated with the establishment of the high pressure behind the cold front than with immediate postfrontal conditions. Thrushes from southern Sweden head southeast in the immediate post-frontal weather where winds are strong,[3] but those departing after winds have abated orient southwest. These may be separate populations with different intended tracks making use of suitable winds. Short distance diurnal passerine migrants tend to depart in anticyclonic weather, or in the slack areas of low pressure where cloud is breaking and winds are light e.g. in slow-moving occluded depressions in the final stages of their lives.

Migrant Scandinavian birds that winter in Britain exploit the tail winds associated with an anticyclonic situation. Massive arrivals of thrushes, finches and Goldcrests take place in northeast to east winds which hasten the crossing of the North Sea. Larger falls than normal occur if the species has had a good breeding season, and we shall see later how some non-oriented species disperse in a dramatic manner, with their destination controlled by wind direction amongst other variables.

Lest it be thought that migrants must have visible astronomical cues in order to set out on successful migration, there are many examples of birds departing and apparently orienting successfully under overcast skies. In these instances, one or other of the alternative methods of navigation may be implicated. The late Professor William Keeton[77] suggested that learning may enable birds to navigate accurately with less information. Overland, visual cues such as coastlines and other topographical features may assist, and certainly the 'leading-line' method of navigation (where migrants follow such features when the trend is in the general direction of their heading) is a well-known and proven phenomenon. Since leading line behaviour is most noticeable in conditions in which drift is highly likely, such as strong adverse winds, it has been attributed to the need to assess and overcome the drift. Many diurnal migrants use this method, and therefore tend to be less prone to drift than nocturnal migrants, though there is evidence that some of the latter may also follow this strategy. On a fine night, coastlines are visible enough to act as a leading line and it has been suggested that reflected moonlight assists migrating waterbirds by revealing the location of suitable waters. In calms or headwinds when ground speeds are low, Woodpigeons depart from southern Sweden in narrow corridors influenced by topographical features. They are loth to cross the sea, and use the shortest route.[4]

Well-oriented songbird migration over Cape Cod, USA, takes place regularly under fully overcast skies and over fog, and waders orient accurately under many unfavourable conditions, including heavy rain. In northeast England, radar observations showed waders departing in the correct direction under total overcast, though there was no evidence to suggest that they had not, in fact, reached the cloud tops. Most departures, on both feeding and migratory movements, were in strong following winds.[48] Passerine night migrants over the same area have also been observed to depart under full overcast, though in almost all cases the weather was anticyclonic with light winds, and the cloud base was low. This suggests that clear skies existed above a subsidence inversion, with navigation easily achieved once the low cloud top was reached. Diurnal Chaffinch migration often remains strong under overcast skies, particularly in semi-constant winds, though this is possibly more common later in the season when urgency is greater and the birds move in less than optimal weather.

Although radar has produced conflicting results in so far as migration in or under cloud is concerned, problems in interpretation may well arise due to a poor knowledge of the complete cloud structure above the observation site. It is notoriously difficult in some instances for an observer on the ground to assess the depth and height of cloud layers, even with all the instrumental meteorological data at his disposal – and I include radar and satellite data.

Just as optimum weather enables migrants to set out successfully, the presence of dense cloud, rain, poor visibility and strong winds from directions opposed to their heading forces them to remain on the ground. As we have seen, the degree to which migration is delayed depends partly upon the species, partly upon the physiological state of the individual, and partly upon the intensity of the adverse weather. This weather operates against successful navigation and speedy migration in the preferred direction. The circumstances under which delays occur are most frequently found in frontal conditions. Referring to the typical approaching depression (see Chapter 1), it can be seen that in autumn, thick cloud, poor visibility in precipitation, and strong headwinds or crosswinds invariably occur ahead of and within a warm front or occlusion. The degree of persistence is related to the speed and activity of such a front, and the depth of the associated depression. The more active fronts have the more intense precipitation and thicker cloud, while a slow moving front results in more prolonged poor weather over a particular region. The depth of the depression influences the strength of the wind. Needless to say, these are generalities, since the precise detail varies from front to front and from depression to depression. Notwithstanding the earlier paragraphs on migration under overcast, an approaching active depression grounds migrants, or delays their initial departure until the clearing weather to its rear passes over the area. The lack of favourable migration weather over a long period results in abnormally late records of summer visitors and delayed passage migrants. Such an autumn in Britain was that of 1954, which was exceptionally windy with frequent southwesterly gales, and the accompanying mild weather ensured a continued supply of insect food to assist survival.

The persistence of blocking anticyclones assists migrants by providing optimum weather for a considerable time, but, just as likely, the wrong position of such a block will result in the deceleration of fronts over just the area from which migrants are to depart. The western or southwestern flank of a block is continually beset by depressions which are forced to turn and slow down. It is not unusual for a sequence of depressions to pile up against a blocking high, with successive occluded fronts becoming slow moving over a particular region, preceded by a more or less continuous succession of strong adverse winds, belts of rain and cloud.

SPRING MIGRATION

In spring – with a greater urgency to reach the destination, and with the worst weather likely at the end of the journey rather than at the beginning – the meteorological factors associated with migration are in some respects different to those in autumn. The migrants must ensure that they do not arrive at successive latitudes too early for a satisfactory food supply, and several arctic species which risk this, notably waders and geese, break their migration by using traditional staging posts where they renew fat reserves. The need for a rapid return from winter quarters often results in movement being in more disturbed weather than in autumn, but in general, warm airflows from the south are associated with northward passage in the same manner as are cold winds for southward passage. Summer visitors tend to arrive rapidly and unnoticed until taking up territories on their breeding grounds.

Because of the steadily improving weather conditions in spring, late arriving migrant species are frequently more constant in their date of arrival than the early migrants, since the latter are more likely to be delayed by poor weather. As the European anticyclone is well-established by April (see Fig. 48), migrants following a trans-continental route in spring tend to be more regular and to proceed at a faster rate than those using the western seaboard. This is so with the Willow Warbler, whose western populations are more affected by frontal systems approaching from the Atlantic. In fine weather and following winds, large numbers of migrants have been seen on radar overflying the coast of southern England to continue inland. Few birds appear at coastal observatories at these times.

Warm winds from a southerly point are a feature of the western flank of anticyclones and ridges, often ahead of warm fronts. With increasing solar radiation, clear or broken skies in high pressure zones and warm sectors are invariably associated with rising temperatures. The long distance oceanic crossings of autumn are not repeated on such a large scale. Over the eastern Atlantic, migrants bound for Iceland and Greenland take shorter, more direct routes. Goose emigration from northwest Scotland and the Outer Hebrides reaches its greatest intensity in anticyclonic weather.[189] The optimum pressure pattern appears to be a polar high centred over the Faeroes area, with good visibility and a following southeasterly wind on its western flank. Similar synoptic situations are used by geese moving north in North America (and perhaps, also, arctic waders). They tend to migrate in clear weather with warm following south to southwest winds. This allows maximum energy saving by minimising energy expenditure (via optimum wind conditions), thermal stress (temperature), and displacement (ideal navigation conditions) all vital to ensure available energy reserves on arrival[13] (see Chapter 5). Flight altitudes are doubtless lower than in autumn, since there are not the strong following winds aloft. As we saw earlier, the jet stream in the upper warm air ahead of an eastwards moving depression is from a northwesterly point, and thus low level southeast winds veer gradually with altitude, and high flying migrants would meet headwinds.

Radar watches in the Outer Hebrides have shown that all sizeable movements of passerines coincide with light or following winds.[91] Waves of migrants further south in Britain, and in northern Europe, often arrive in favourable winds and warm weather, but also in headwinds.[85] There appears to be more inhibition, however, in strong headwinds or crosswinds than there is stimulation by tail winds. Over the North Sea, waders and passerines often fly against the wind. As in autumn, sudden movements occur when favourable weather succeeds a spell of poor weather. The density of passage in the eastern USA has been significantly correlated with high or rising temperatures, low or falling pressure, a low but rising humidity and an onshore (southeast) component in the wind direction.[125] Passage is strong in the west central region of a high pressure cell, but declines on the arrival of humid tropical air. From this, one can infer that migration occurs chiefly in the warmer south to southeast airflows on the western flank of a receding high pressure zone, and ahead of a warm front, though the precise factors to which the birds respond are still uncertain. Dense migration also takes place in following winds of warm sectors. Substantial breaks appear overland in shallow warm sector cloud, and frequently the warmest air is found just preceding the cold front – another region in which large movements occur. Northeast to east-bound migrants also depart to the rear of cold fronts, but invariably where these are succeeded by southwest to west winds, and are thus closely associated with tail winds. In Sweden, thrushes with an ENE heading are more likely to migrate behind cold fronts than those oriented north. The delay shown by thrushes moving ENE over Scandinavia is thought to be linked to food availability. Departure behind a frontal system allows the warm airmass to precede the migrants, making invertebrate food more accessible on arrival.[3]

Differences in route between spring and autumn can be found in several instances. Known as loop migration, birds performing this strategy are able to take advantage of suitable seasonal wind regimes, as well as avoid adverse ecological factors. Off eastern USA, little or no northward oceanic passage takes place in spring. Coastal migrants have the advantage of being able to land if meeting adverse weather. Additionally, April and May are the most cloud-free months along the eastern seaboard. In April, relatively cloud-free conditions extend as far north as Nova Scotia, but it is the lack of strong following winds equivalent to those behind an autumn cold front that is the ultimate reason for overland movement. Such a strategy using seasonal winds is also found in those birds (mainly falcons, bee-eaters and cuckoos) which use the monsoon winds between India and northeast Africa. From late October, downwind flight is assisted by the northeast monsoon. Observations of a return on the southwest monsoon are sparser, and the onset of this windflow across the Arabian Sea is rather late for the spring migration – normally not reaching India until June.[107]

ALTITUDE

Two factors determining the altitude of migration have already been mentioned, namely wind speed and direction, but others may be involved, particularly cloud thickness, moisture content and other, non-meteorological, variables. The method by which a bird detects its altitude is generally assumed to be by the use of the ear in sensing atmospheric pressure changes (see also Chapter 1), but other possibilities include the use of flight calls, either by producing a Doppler effect through echoes from the ground, or by communicating optimum flight levels to other migrants in the flock. Altitude poses little physiological problem with regard to breathing; ventilation and gas exchange between lungs and blood are far more efficient in high-flying birds than in any other vertebrate. Reduced air density at higher levels also allows faster flight.

The numerous records of high altitude flight indicate that temperature itself has little effect if flight is in clear air. Lapwings have been recorded at temperatures of −10°C, Starlings at −12°C and Whooper Swans at −48°C. It has been suggested, however, that migrants might fly below the freezing level when in cloud. If so, this would probably depend on the water content of the cloud. In thick cloud with a high water content, as in an active front or deep and vigorous convection, supercooled water above the freezing level may well induce considerable icing on the plumage (see Chapter 11). Birds may therefore select an altitude at which they can reach an equilibrium between metabolism and the humidity and temperature of the environment, since evaporation from the skin and by respiration could be considerable in dry air.

The higher the temperature of a mass of air, the more water vapour it is able to hold. Thus, at a constant relative humidity and atmospheric pressure, the rate of water loss of a bird by evaporation decreases with air temperature. At the heights at which passerines normally migrate in northern latitudes, water loss is negligible. There can, nevertheless, be marked changes in temperature and humidity with altitude. In Fig. 26 the amount of water (in vapour and liquid form) required to saturate the air is shown for various altitudes in typical airmasses which an autumn migrant travelling from northern Europe to West Africa might encounter. The greater the value along the horizontal axis, the more moisture a migrant would lose. It can be seen that the altitude at which it flies becomes more important at lower latitudes. For example, flight at 700 m would give minimal loss until it reached the Sahara, where low level heat and very dry air would cause severe dehydration. At 1500 m dehydration increases over Iberia in the subsiding air above a marked inversion, and the contrast between the low level air over the desert and that over the sea in the Trade wind zone is considerable. Even though the size of a temperature inversion may be just as great in northern Europe (as it was in the polar anticyclonic air illustrated in Fig. 26) the moisture loss is far less important, since temperatures are considerably lower. Therefore, with regard to moisture loss, only where temperatures are high and humidities are low at low levels is it beneficial for a migrant to climb to a higher altitude, although it is still uncertain to what degree moisture content of the air influences the altitude and range of long distance migrants.

It has been determined by radar that small migrants do indeed climb to considerable heights in subtropical airmasses, often above 2 km and sometimes as high as 6 km.[107,146] Autumn passage over the Sahara is at altitudes as high as 3 km, where migrants avoid the sandstorms which can be lethal to lower flying birds. For this reason, many species are hardly ever seen on the southern shores of the Mediterranean. Similar altitudes are used in spring, when favourable southwest winds are present over the desert at 2 km. In the colder airmasses over North America and Europe, most small passerines fly lower than 2 km, often only at 250 to 500 m at night.

Studies of flight altitude have revealed that many migrants fly at the height of the most favourable winds because in this way their flight time is reduced. This is especially important if winds change with height. I have already

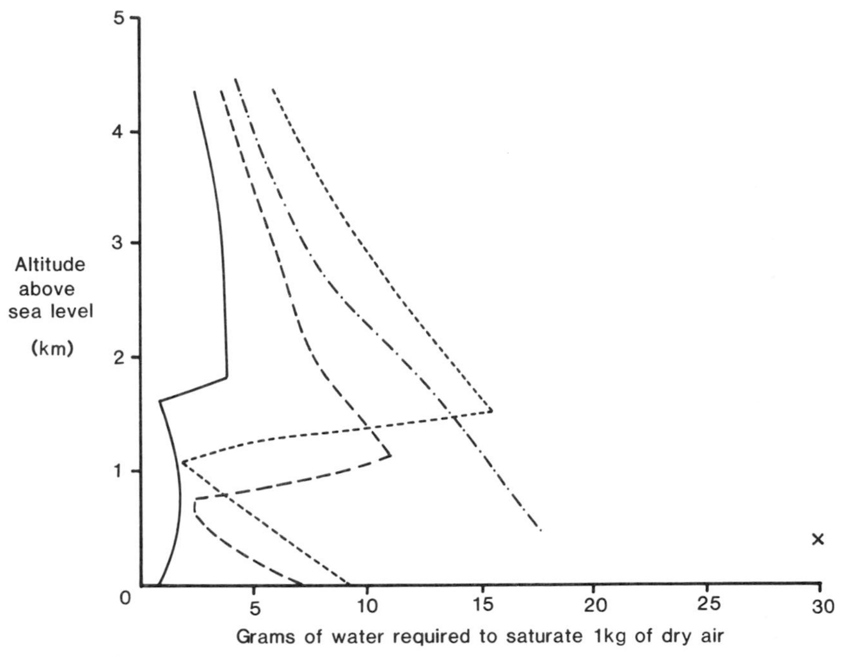

Fig. 26 Amount of moisture required to saturate air in typical airmasses encountered by a migrant flying from northern Europe to West Africa in autumn.
________ Polar airmass in anticyclone, NW Europe (56N)
– – – – – – Tropical maritime airmass, western Mediterranean (36N)
.... Trade wind airmass off west Africa (16N)
– · – · – · – Tropical continental air over Sahara
× Typical maximum surface value, central Sahara
High values indicate dry air resulting in high water loss from a migrant.

shown examples in which wildfowl use or avoid strong upper winds, depending on their situation. Autumn migrants departing immediately behind a southward moving cold front benefit by remaining at low levels, however, since the low-level following winds back with height to become strong crosswinds aloft where the jet stream is more or less parallel to the surface front (Fig. 3). Short distance migrants tend to fly at low levels in any case, as the energy used in climbing is wasteful. Except for many waders, long non-stop flights are not generally performed over ecologically hospitable areas, but only across barriers such as deserts, oceans and high mountain ranges. These species have markedly higher fat levels to carry them successfully over such barriers.

Short flights show a high wind selectivity – the rapid wind changes experienced at low levels permit movement only when the wind is favourable.[4] Most species tend to fly higher in favourable winds and clear skies, but at low levels in strong headwinds, low cloud base and heavy precipitation – if indeed they fly at all. As already mentioned, observations by radar often conflict with those by visual means, since the radar cannot see below its horizon, and the latter cannot detect flight at higher altitudes. Therefore problems arise with interpretation. However, investigations using radio-telemetry confirm such wind selectivity, showing that migrants may continually assess conditions in their immediate air-space and adjust their altitude, heading and speed accordingly.[200] Low-flying migrants tend to use topographical features to a greater degree. Finches follow coastlines more and fly progressively lower as the windspeed and headwind component increase.[121] Chaffinches bound for Britain in autumn fly high across the North Sea with an easterly tail wind, but in westerly headwinds fly low along the continental coast and cross into southern England by way of the Pas de Calais. Differences in altitude of migrating Chaffinches in Holland have been related to the sea breeze effect (see Chapter 4). The need for a constant wind direction constrains birds flying along the coast to climb above the sea breeze, whereas those inland remain at a constant altitude.

Selection of favourable winds varies among species, and in southern Sweden, Woodpigeons and Jackdaws are more selective than finches. In Chapter 2 I showed the benefit of flying high in a tail wind to avoid low level turbulence and the risk of stalling in gusty winds, and migration studies tend to confirm this. Over southern Sweden in autumn, the diurnal low-level migration of passerines takes place at a higher altitude when winds are favourable – the most intense passage being in northeasterlies. With headwinds from between west and southeast, passage is small. For Woodpigeons, topographical features assume a greater importance with light headwinds. The leading-line effect attains a maximum at low levels for many species migrating against headwinds,[4] and in montane regions, such as the Alps, funnelling of small migrants through valleys and passes is magnified in poor weather.

SUMMARY

Migration is a seasonal movement originally evolved to avoid food scarcity in the natal area. In autumn it is stimulated at the time of peak migratory urge by conditions which favour rapid passage and accurate navigation, although many birds can apparently orient successfully under overcast skies by using methods other than those relying on the sight of astronomical cues. The clear skies and light winds created by subsiding air in anticyclonic situations stimulate the inception of migration, while stronger winds in a favourable direction in polar airmasses both stimulate and hasten passage – particularly for those species which must undergo a long distance oceanic crossing.

In spring, similar conditions apply, but passage is more rapid and urgent. Some species use a different route in autumn to make use of seasonal wind regimes.

The prevailing wind (the mean airflow over a specific area) is often opposed to the normal direction of migration, especially in autumn (see Fig. 50). Those migrants which are most influenced by weather when crossing the mid and high latitude depression tracks are thus opportunists in the timing of their movements, and avail themselves of those relatively few situations that interrupt the prevailing flow.

The altitude of bird migration is thought to be determined, among other factors, by atmospheric moisture content and windflow. The former controls metabolic processes with regard to evaporation of body fluids, while the latter controls speed of migration and drift – a subject to which I shall turn in the next chapter.

CHAPTER EIGHT

Migrational drift and displacement

Once migration has begun, a bird may encounter many variations in the weather, and responses from species and individuals differ considerably. A migrant cannot predict the conditions that it will meet on its journey, although, as we have seen, it chooses to depart in favourable weather offering at least a degree of persistence.

ORIENTATION AND COMPENSATION

Weather that inhibits successful navigation and continued onward passage is not infrequently encountered, and one of the more serious problems arising when orientation is hindered is that of drift by the wind. Many arguments have arisen over the degree to which a migrant can compensate for drift, or subsequently re-orientate. A certain amount of drift must be expected, of course, but the extent of it will depend on the particular bird – its size, weight and power. Strong winds seem generally to have a less deflecting influence on heavier and faster birds. Conflicting interpretations have been placed upon radar and visual observations, but the fact remains that there are certain synoptic situations which result in large falls of migrants at particular locations when disorientation or drift is presumed to have occurred. The absence of a suitable landing place for landbirds when disorientated over the sea means that they are somewhat at the mercy of the winds. Displacement by wind from the normal migration route varies among individuals and species. Experienced adults may be less prone, and more able to compensate (or better able to judge suitable departure weather) than immature birds, resulting in a higher proportion of the latter among grounded birds in poor weather. It has been suggested, however, that displacement may not be passive, but is shown by pioneers taking advantage of wind to explore beyond their normal range at minimum cost. It has also been proposed that birds hardly ever become lost, and that vagrants are merely individuals (mainly immatures) undertaking long distance exploration.[199]

Several studies have indicated that high flying migrants may not always allow for lateral displacement, so that crosswinds give extensive drift over land and sea. Low flying birds, on the other hand, are able to correct for drift by using landmarks, and possibly seamarks. However, since populations with different intended tracks migrate in varying proportions in different wind conditions, movements shown by radar and visual observations may give a picture that is difficult to interpret. They may show apparent, but not real, drift – known as pseudodrift. Some birds may try to correct for drift over the sea by using wave movement, which is parallel to the surface wind direction. Nevertheless, drift may still occur, since there is no precise relationship between the wind at the birds' altitude and that at the sea surface. Even at an altitude of a few hundred metres, the change in wind direction and speed can be quite substantial, depending on wind speed and temperature lapse rate among other factors. In most synoptic conditions in which migration takes place, however, these changes are minimal.

The potential for displacement is greatest when a migrant is completely disorientated, in which event it lands or attempts to compensate by following ground marks. Where it cannot do either, the track of the bird, although remaining in the same direction for a time, becomes irregular with frequent changes in direction and eventually a downwind movement. Disorientation affects birds in a way that is similar to its effect on human pilots, with the added problem of being unable to apply reasoning to the situation; the effects are compounded by inexperience. There is evidence, however, that when visual cues are absent, migrants are able to detect wind direction and actively orient downwind.[196,200]

Disorientation in light and variable winds, or in calms, with haphazard flight over the sea, leads to progressive dwindling of fat reserves and finally total exhaustion, but downwind flight in a definite direction at least increases the distance flown for the same expenditure of energy, and increases the chance of a landfall. Light-weight birds that make landfall in these circumstances may have used all fat reserves, and even made inroads into other tissues. Many do not recover and are doomed. Subsequent re-orientation may nonetheless be undertaken both by adults and immatures should conditions improve. Radar observations and recoveries of ringed vagrants after return to their normal range have confirmed this to a small extent, but the degree to which it can be achieved is not yet clear. Reorientation of migrants has been recorded on both sides of the Atlantic. In the northern and western islands of Scotland, thrushes and other small night migrants from northern Europe re-orient from a westerly track to a south or southeasterly track at dawn when finding themselves drifted in easterlies over the adjacent seas. It has also been shown by birds migrating off eastern North America, particularly late in the night when winds are unfavourable for continued southwestward flight.

The remarks made in the previous chapter on the difficulties of assessing cloud structure in migration analyses apply also to observations of drift. Radar tracked migrants have been seen flying straight courses when apparently within or between opaque cloud layers. They showed a strong tendency to move downwind but corrections were consistently applied to bring their track closer to the preferred direction. Whether the corrections were assisted by the sight of astronomical cues through cloud breaks undetected by the observers, or by other navigational aids, is not known. The effect of fog and mist is similarly difficult to assess on some occasions. The depth of fog can be such that, even though only a few hundred metres, the sky or cloud above is obscured. Poor visibility in frontal situations is likely to be associated with precipitation and thick cloud layers aloft but, below a subsidence inversion, fog or low cloud over land and sea is more likely to be topped by clear skies, and migration can continue unimpeded above it. Thus one cannot always assume that fog, mist or cloud seen from the ground has any disorientating influence upon migrants.

In North America, Herring Gulls carrying miniature radios were released away from their home area and tracked. In good visibility, birds homed from 50 km using the most convenient route, but when fog covered the release point, they undertook a systematic search until the fog edge or a familiar landmark was reached, thence homing satisfactorily.

DISPLACEMENT IN AUTUMN

Western Palearctic migrants

The cyclonic approach of autumn migrants from Greenland and Iceland southeast towards Britain frequently has a trajectory in which downwind flight brings them into a progressively backing airstream, so that the approach to north and west Britain is from the west or even southwest. Some compensation for drift is probably made, but the resultant track in strong winds may not be completely in accordance with their preferred direction. If frontal conditions are encountered, drift can become excessive and downwind. This is particularly disastrous for those birds migrating through a col or ridge, rather than with northwesterlies, since they may meet the strengthening low level southerly or southeasterly headwinds associated with a frontal system on the western periphery. Drift is then out to sea. Presumed

Icelandic Redwings approaching the Outer Hebrides from the NNW have been seen on radar to turn in a stream to avoid rain clouds, and others made no headway against a southerly gale. Yet others were disorientated by cloud on an occlusion, to be blown downwind in a northward direction.[91]

There are innumerable occasions in which birds have landed on ships; Greenland and Iceland birds are regular visitors to ocean weather ships west of Britain, while Icelandic and north Norwegian birds visit ships in the Norwegian Sea. Many are exhausted, particularly the drifted birds, but these may nevertheless be only a very small fraction of the total on migration. The potential for recuperation of these ship-borne migrants may yet be high. One exhausted first-year Greenland Wheatear recovered rapidly when fed by the crew of a weathership stationed 450 km south of Iceland.

Like the smaller migrants, geese and swans are prone to drift when migrating to Britain from the northwest or north. In autumn 1963, Pink-footed Geese driven from Iceland by early heavy snowfall penetrated strong winds to the north of Britain and were drifted east of Scotland to southern Norway and Denmark, where Icelandic birds are not normally seen (Fig. 27). A large number re-oriented, and were observed on 26th September approaching eastern Scotland low over the sea against westerly gales. Spectacular arrivals were noted in this region, but southward movements at the same time over the Highlands suggested that at least some approached from a more normal direction – perhaps birds originally on a slightly different heading, or having departed from Iceland somewhat later.

Further east, wildfowl migration in the Baltic is held up by low cloud and rain, yet fog over southern Sweden apparently serves only to deflect Eiders to cross the land rather than fly along the coast. Flocks of Long-tailed Duck, Common Scoter, Brent and Barnacle Geese migrating between the White Sea and the Baltic avoid passage in winds liable to cause a strong deflection, since they are apparently unable to compensate for drift. With onshore winds, ducks tend to fly inshore, but when winds blow away from the land, most movements take place far out to sea.

The number and species composition of displaced European migrants appearing in Britain depend upon the size of the migratory population, its heading, the time of year in relation to the migrants' physiological state, the severity and persistence of the prevailing weather type and the axis of any area of poor weather, although some of these factors assume a greater importance than others. The weather conditions over the North Sea affect visible migration down the whole east coast of Britain more than any other single factor, but the correlation between weather and falls of migrants is by no means simple. Wind directions most commonly associated with falls of North Sea migrants – whether waders in late summer or passerines in autumn – are those with an easterly component. Very few occur in westerly winds. The situation combining anticyclonic weather over Scandinavia and strengthening easterly or southeasterly winds ahead of a warm front or occlusion approaching the North Sea from the south or southwest, is one which is familiar to ornithologists eagerly awaiting falls of night migrants in autumn. Under these circumstances, the thickening frontal cloud with rain, the strengthening of crosswinds or headwinds, and finally poor visibility just ahead of the surface front, exhausts small migrants in large numbers.

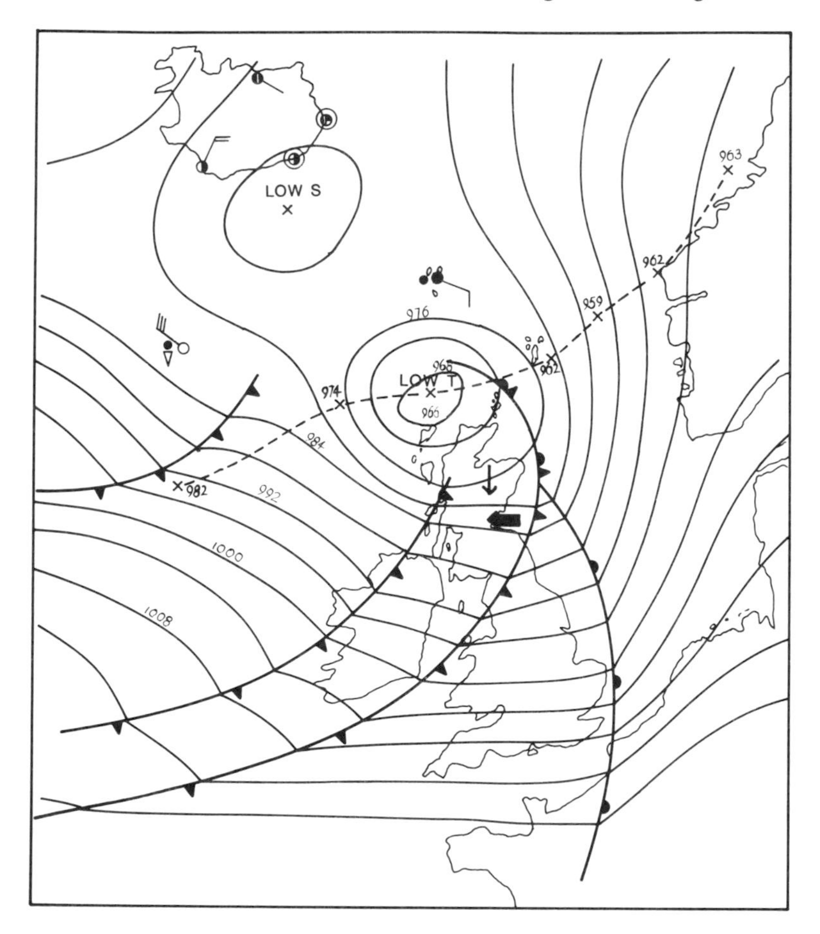

Fig. 27 Drift of Pink-footed Geese, 26th September, 1963, 0600 hours. Track of Low T is shown at six-hourly intervals with central pressure marked.
← arrivals on 26th September p.m.
🡄 arrivals from 26th September p.m. to 27th September a.m.
As Low T approached NW Scotland, Low S filled. Previous strong winds over Iceland abated and cloud broke, inducing geese to depart from very cold conditions and extensive snow cover. Those departing on a southerly track probably flew into the NW winds behind Low T, but others would have run into easterlies to the north of Low T and then swept down to its rear into the North Sea when at its most intense. Many landed in Denmark and S. Norway behind the occluding front which reached that area in the afternoon of the 26th. Others reoriented in gale force westerlies. These gales covered the whole of the North Sea, concentrating seabirds into the German Bight (see Chapter 12).

Individuals of all ages are affected, but more especially the inexperienced birds of the year. Again, exhausted migrants land on ships, and, more recently, have found doubtful refuge on gas and oil platforms. Most of these falls coincide with mist or fog, and even in western Britain, arrivals of night migrants are closely correlated with extremely poor visibility. The majority of birds in west coast falls are of British origin, however, since the onset of autumn migration from Britain is begun in fine weather, and the migrants are therefore unlikely to have been subjected to much displacement after encountering adverse weather.

Whatever the interpretation of apparent drift – whether passive displacement or active flight downwind – the location of the east coast sites that experience groundings of migrants depends upon their position in relation to the frontal weather. SSW-oriented migrants from Scandinavia (and there are many such species) overfly the southeast corner of England as well as the continental coastline and the North Sea itself, but in fine weather remain high and out of sight, especially the long distance migrants. It is only in deteriorating weather or strong crosswinds that such migrants are observed off course and grounded. Frontal displacement and disorientation usually result in localised groundings – shortlived at any one site if the frontal weather is moving. Adverse winds drifting birds into coastal areas frequently change direction on the passage of the front, so that the fall ceases. At the same time, arrivals continue on the forward side of the front, with the arrival point of the migrants progressing with the front. The species composition of the falls may also change with time and place.

A remarkable example of a fall of migrants took place on 3rd September, 1965 on the coast of East Anglia, when an estimated half a million birds landed along 40 km of Suffolk coastline alone.[31] The synoptic situation embraced all the conditions conducive to massive falls. The month began with a building ridge of high pressure moving south over Scandinavia, so that by 2nd September ideal conditions existed for migration from that region. Over the southern North Sea and English east coast, a fresh showery northeasterly prevailed, and small arrivals of passerine night migrants were observed in coastal areas. The northeasterlies were part of the circulation of a depression over the Alps, which moved in an unusual direction, northwest towards Britain, on 2nd September. A warm sector was carried westwards round its northern flank, preceded by strong north to northeast winds, extensive cloud and rain from southern England to Denmark. Over southern Norway and Sweden, the ridge persisted. By dawn on 3rd September the frontal situation had become complex, with the thick frontal cloud south of a line from the Humber to Denmark. To the south of this the depression had also become complex between northern France and north Germany. By midday the main low centre was positioned in the southern North Sea, with another small centre over southeast England. Considerable numbers of migrants were arriving by then between Yorkshire and Norfolk, but it was further south that the extraordinarily concentrated falls took place a little later. During the afternoon, the southern centre deepened and moved northeast, while the northern low moved northwest (Fig. 28). Over the coast of East Anglia, winds backed temporarily from NNW to SE ahead of the southern low before veering to west to northwest behind it in the late evening. It was this backing to a crosswind, accompanied by heavy rain, that brought huge numbers of migrants tumbling from the sky in the early afternoon – in some places only a few minutes after the wind change.

The immensity of the fall reflected the intensity of the poor weather over the main North Sea passage route. The SSW-oriented migrants had encountered the frontal cloud over the sea, some in tail winds north of the depression, while others penetrated the area of variable winds or even perhaps the southwesterlies to the south. The brief change of wind ahead of the second low centre, following a reduction in altitude, led to a successful landfall. Since many were exhausted and of low weight – with large numbers washed up dead on the tideline – the random flight in heavy rain and

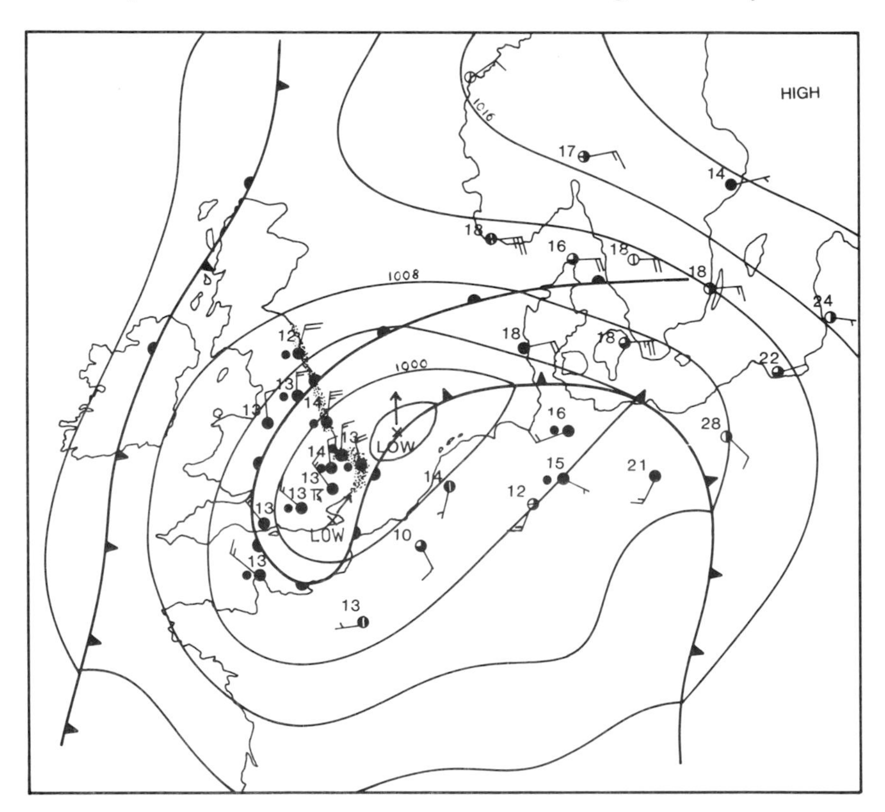

Fig. 28 Autumn Migrant fall, 3rd September, 1965, 1200 hours. Arrows show subsequent movement of low centres. Temperatures in °C. Stippling marks English coast affected by falls. Note southern limit near Orford Ness, Suffolk, corresponding to onshore windflow.

headwinds had clearly been very debilitating for some.

The sheer magnitude of the fall was overwhelming. One experienced observer, on a 4 km walk along the shore early on 4th September, estimated 15,000 Redstarts, 8,000 Wheatears, 4,000 Pied Flycatchers, 3,000 Garden Warblers, 1,500 Whinchats, 1,500 Tree Pipits, 1,000 Willow Warblers and 500 Whitethroats. Redstarts were so numerous that they descended from the skies to perch on the shoulders of town-dwellers. The fall was shortlived, and with breaking cloud and veering winds, the majority of the migrants moved on within a few days.

Massive falls have occurred elsewhere from time to time, and invariably highlight the enormous numbers of migrants which normally fly over unseen. Lighthouses are notorious for causing fatalities when migrants become confused by the light in mist and low cloud, and strike the unlit dome and tower. In Britain, one of the most infamous lighthouses was that on Bardsey Island in North Wales. On one August night in 1968, at least 560 warblers were killed, including 111 Grasshopper Warblers.

While frontal weather results in large but localised falls, falls taking place in anticyclonic weather occur on a broad front, and sometimes over several days. These embrace successive waves of migrants departing in the persistent fine weather over their area of origin. Whereas the massive falls described above include birds of all ages, most of the birds in these fine weather falls tend to be juveniles. Many are thought to be on non-oriented dispersal rather than merely drifting from a defined migration route. In Britain, the area affected by the falls is dependent upon the position of the anticyclone, with range and direction of dispersal guided by the wind. Arrivals on the east coast occur when the high is over northern Europe, while a high further south or southwest brings birds to the south coast or Irish Sea area. These falls are augmented after a successful breeding season.

Some species appear regularly at east coast sites in these anticyclonic situations, often in considerable numbers. In October 1951, a fall of Robins was observed along the whole of the east coast in light to moderate easterlies. This wind regime lay south of an anticyclone over Scandinavia which also ridged into Russia. Several thousand birds appeared between Shetland and Kent – mostly from southern Scotland to Norfolk, with some reaching the west coast. Although analysis at the time showed overcast conditions with fog banks over the North Sea and Baltic, it seems likely that the fall was of non-oriented individuals, since in the north, birds arrived in clear weather, and in the absence of fronts further south, the cloud was shallow. The subsequent recovery of ringed birds in the Mediterranean area, however, indicated that at least some oriented birds were drifted in the easterlies and had re-oriented successfully to their original destination – a destination usually that of Robins from eastern Europe rather than Scandinavia. The fall had clearly included birds of a more distant origin than those normally visiting Britain.[191]

The Blackcap is another migrant whose arrivals often occur on a broad front. Falls in recent years have taken place in a variety of synoptic situations, but the features that all have in common are winds from between northeast and southeast, with high pressure between northern Germany and Scandinavia and depressions remaining on a southern track – a similar situation to that of the 1951 Robin 'rush'. Much of this Blackcap passage is of Scandinavian birds, where western populations are SSW-oriented.

Yet another regular migrant which uses anticyclonic easterlies is the tiny Goldcrest. This species varies in numbers, and appeared to have an excellent breeding season in 1975. Huge numbers were seen in Finland in late September, and on the 10th and 11th October, thousands arrived on the east coast of Britain between Yorkshire and Kent. It became clear that their movement was aided by the extension of a ridge over Scandinavia from an anticyclone over northern Britain. To the south of this ridge, fresh north-easterlies prevailed from the Baltic to Britain.

Other small landbirds arrive in rather different circumstances. The fact that all autumn Barred Warblers on Fair Isle were juveniles of a species that migrates SE from northern Europe, led Kenneth Williamson[185] to postulate that arrivals were perhaps due to post-juvenile dispersal in anticyclonic weather, followed by drift in southeasterlies to Britain, or a prolongation of the dispersal if the fine weather extended over the North Sea. Another SE-oriented migrant, the Red-breasted Flycatcher, shows a pattern of arrival also suggestive of post-juvenile dispersal, although a sprinkling of adults in autumn may indicate the drift of individuals from a small SW-oriented stock from more western breeding areas. On the other hand, Ian Nisbet[123] found a significant correlation between arrivals of Barred Warblers and Red-breasted Flycatchers at Fair Isle and a combination of high temperatures and light winds with a southerly component over Germany, and maintained that the birds were on reversed passage. The most common situation to produce falls was that of an anticyclone over eastern Europe, with southeast winds on its southwestern flank – especially if the high extended into Scandinavia. The greater frequency of Barred Warblers on Fair Isle than further south supports this hypothesis of an oriented northwest movement rather than downwind drift, though they sometimes arrive in southeasterlies along with 'drift' migrants. Overshooting on reversed passage must necessarily result in some mortality, although arrivals on west to northwest winds at Fair Isle may be birds on redetermined passage from islands further north.

Reversed movement in autumn is undertaken by several species, usually initiated by rising temperatures and winds opposed to their normal heading. SE-oriented Lesser Whitethroats, and Blackcaps from east of their migratory divide along longitude 12°E, also make reversed movements to eastern Britain. Ringing recoveries have illustrated these. A female Blackcap ringed in the Netherlands in September, 1973 flew the 'wrong' way in col conditions, to be recovered the next day 1,000 km to the northwest in Shetland.

West to northwest reversed passage is also undertaken by Yellow-browed Warblers. Among other situations (which included the northern edge of the Robin 'rush' of October 1951), they appear to circumnavigate the northern flanks of depressions to reach Britain and Ireland from southern Scandinavia on winds between north and east. This widespread Asiatic species has become more abundant in recent years, particularly in 1975 and the early 1980s, with arrivals often coinciding with the more extreme rarities from Central Asia (see Chapter 9).[198] The increase may be the result of an as yet undiscovered population explosion.

Further south, Melodious Warblers arrive in southern Britain more in fine anticyclonic weather than with frontal conditions, and are often associated with southeast winds. They may also be on reversed passage from France or Iberia though, together with Icterine Warblers, post-juvenile dispersal is an alternative answer.

Besides the species most abundant in the September 1965 fall described earlier, there were numbers of a large variety of other small birds and waders. Though the majority were of SSW-oriented summer visitors to Scandinavia, the fall included some SE-oriented species, especially the Spotted Flycatcher, but also Icterine and Barred Warblers, Lesser Whitethroat, Redbacked Shrike and Red-breasted Flycatcher. A possible explanation for such arrivals is that of post-juvenile dispersal, or drift from a redetermined movement of those birds which had earlier undertaken reversed northwestwards passage towards Scandinavia. The latter is possibly the more likely, having taken place under the influence of the southeast winds and rising temperatures in the Baltic area (Fig. 28).

Migrants from northern Europe occasionally reach the western seaboard of the British Isles if their route continues to lead them into persistent poor weather – a common enough situation with a blocking anticyclone over southern Scandinavia. Land birds may be drifted out over the ocean by east to southeast winds, or overshoot with a strong tail wind if merely on passage to Britain or France. On the island of North Rona off the extreme northwest corner of Scotland, visual observations show that European night migrants are relatively regular in autumn except in strong winds between west and north. They are commoner than further west on St Kilda, since North Rona lies just within range of one night's flight from southwest Scandinavia. Grounded migrants awaiting good weather on such exposed and barren islands often succumb through lack of cover and food. Valuable fat reserves are used up and cannot be replaced. Factors such as the unavailability of food may in fact induce grounded migrants to depart as soon as possible, often in sub-optimal weather.

Displacement may even take birds well out into the North Atlantic, rarely as far as Iceland, Greenland or North America. Again, European continental migrants appear on weather ships. On one such vessel stationed in the Norwegian Sea, most passerines arrive in fresh south to southeast winds associated with the passage of a depression.

As we have seen, headwinds and rain tend to inhibit migration, but just as overcast skies do not stop the inception of migration, passage can and does continue when migrants meet overcast skies *en route* – particularly the diurnal migrants. Poor visibility and low cloud deter Chaffinches and Woodpigeons from setting out over the sea, but overland migration continues. In southern Sweden, typical Woodpigeon migration is affected by fog, but nevertheless approximate tracks can be maintained without visual ground contact.[4] Woodpigeons departing the Kent coast in late autumn mist and fog make only short flights over the sea before returning, when several hundreds may perch on the cliffs.

Hirundines, which can feed on the wing while migrating, are noted for their infrequent catastrophic appearances when caught on passage by sudden cold spells, during which food becomes very scarce and the birds weaken rapidly. Large scale mortality of Swallows, House Martins and Alpine Swifts took place in the first week of October 1974. The European high pressure zone normally present was completely absent, and instead the block was further west over the Atlantic, allowing a succession of depressions to move SSW over the North Sea with cold northerlies in their wake. Some parts of West Germany had their second coldest October since 1781, while Switzerland experienced its highest October snowfall since 1917. Large numbers of hirundines were trapped by this weather before crossing the Alps, and tens of thousands (mostly House Martins) moved into southeast England from Denmark ahead of southwestward moving fronts. In Kent and Sussex, exhausted birds entered houses and roosted on window ledges and in trees, during a spell which even gave snow in the region. Other insectivores overtaken by cold weather on passage find it necessary to turn to alternative food, and readily consume berries and fruits.

With some important exceptions (see Chapter 10), most migration over the

Mediterranean is on a broad front, with no radar evidence of concentrations of migrants across the narrowest points. Leading lines may nevertheless guide diurnal migrants towards narrow crossings. The incidence of active frontal depressions is less than in northern Europe. Warm fronts are invariably weak, but cold fronts are particularly active – understandable when one considers the temperature contrast when polar air reaches the Mediterranean. From late autumn, depressions are often stormy with heavy rain and thunderstorms, presenting barriers to southbound migrants.[212]

With low pressure over northwest Africa, easterly winds can be strong enough over Iberia to drift migrants westwards. In the absence of easterlies, many more small night migrants cross the Strait of Gibraltar than over the sea to the west, but when easterly winds are accompanied by poor visibility, cloud and rain, grounded migrants are found all along the south coast of Iberia. Moon watches in easterlies have confirmed that migrants are subject to drift. The interpretation of migration in this area has its problems, since easterlies in fine weather frequently blow only at low level, so that birds aloft may well be flying in a light southwest wind. Easterlies also bring haze in otherwise dry and sunny weather, obscuring the opposite coastline at the narrowest points. Some diurnal migrants avoid crossing under these drift conditions. Hirundines follow the coast rather than fly directly across the sea, and erupting Fan-tailed Warblers make landfall in Gibraltar in larger numbers than normal.

Further south again, the period from July to September along the southern border of the Sahara desert is typified by the passage westwards of 'monsoon cyclones'. These are disturbances which form as far east as Sudan – probably initiated by high ground – and move westwards to the north of the Inter-Tropical Convergence Zone in the upper easterly airflow. They intensify steadily and, of those that move out over the sea, a few eventually become tropical storms. Although characterised by cloud clusters and shortlived squalls, it is the easterlies (of up to 20 m/s at 3 km) that pose a problem for small south-bound migrants, especially as at that stage they are tired, light in weight, and therefore more prone to drift. Migrants in the far west are most likely to drift out to sea, and indeed falls have occurred on shipping in overcast and squally weather in September off Mauritania. Some of these birds have been examined and found to be dehydrated, with no trace of fat – an indication of their physiological state by the time they have reached low latitudes.

Conditions in the desert itself are inimical to migrants, and they must cross quickly. Large numbers do occasionally make landfall in oases and the open desert, but not all these migrants succumb, since many have sufficient reserves for onward flight. Swallows are reported dead in the desert more than any other bird, perhaps partly due to their lower flight altitude or their relatively poor thermoregulation, and other small passerines have succumbed in strong winds.[107] Even on reaching tropical Africa, some Palearctic migrants are not immune from disorientation and grounding. In east Africa, some small passerines move south in midwinter from 'staging' areas in the north to their late-winter quarters. The rains associated with the ITCZ typically form low cloud which envelops high ground. Southbound migrants in eastern Kenya are attracted to lights at hill sites, resulting in considerable falls.

Nearctic migrants

The migration of small passerines in North America appears to share the same affinities with the weather as does that in Europe, though the different frequency or structure of weather systems may invoke different responses. Random movements of songbirds watched at radar sites on the eastern seaboard are associated with overcast conditions, rain, fog and stationary fronts. Falls of migrants at the coast accompany such weather, often coincident with strong westerly winds. Unlike migrants in western Europe, however, neither transoceanic migrants nor eastern coastal migrants have the advantage of a landfall when drifted by westerlies. When these winds are strong, coastal birds are drifted out to sea. Many are able to return to the coast only with the expenditure of considerable energy, arriving in an emaciated condition. Oversea passage in post-cold frontal conditions was described in Chapter 7. Many of the late autumn migrants take advantage of favourable winds to cross the western Atlantic in one long hop from Nova Scotia and the northeastern states of the USA to the West Indies and South America. Even small landbirds take part, but are unable to sustain this long passage unless assisted by the wind. Radar has shown large numbers of migrants heading southeast in the northwesterlies,[145] but should the cold front decelerate (as it so often does as the parent depression continues to move away northeastwards with the strongest winds) the migrants overtake the front. Normally the front weakens as it decelerates, with the associated cloud band becoming narrow and broken, with light winds. The migrants can penetrate the front without undue disorientation, and continue their passage towards the tail wind component of the Trade winds further south. Their heading is southeast, but the eventual southwest to west drift in the Trades ensures a landfall rather than a continued flight into the South Atlantic. There is evidence that correction for drift is applied in the early stages of the flight but not latterly.[146]

Although most migrants can penetrate and fly through the frontal cloud without displacement, there are certain conditions which preclude this. Cold fronts which slow down and align themselves NE–SW along the edge of the Gulf Stream current often undergo renewed activity. In this position, with polar air to the north and tropical air to the south, the frontal zone retains its identity as the airmasses overlie cold and warm seas on either side of the narrow but marked sea temperature contrast. Sea temperature changes of up to 10°C have been recorded within 20 km. The vigour imparted to the zone induces frontal waves, which then run along the frontal surface in the upper wind flow, often developing into new depressions. As described in

Chapter 1, a frontal wave thickens the cloud, intensifies any precipitation, and strengthens the winds in the warm air. Thus migrants meeting an intensifying front are liable to be disorientated and come under the influence of the strengthening southwest winds in the warm airmass, whence they are drifted eastwards towards the open ocean. Radar has also shown birds flying slowly and randomly over the western Atlantic in the vicinity of depression centres.

The long southward passage over the westernmost Atlantic between the two Americas is undertaken by wildfowl, waders and some passerines – the most notable of the last being the Blackpoll Warbler. The lateness of this passerine migration, chiefly in October, gives distinct advantages.[44] The mean position of the polar jet is at its northern limit over the Gulf of St Lawrence, meaning that fronts south of this tend to be less active, and the formation of wave depressions on the migration route has still to reach its winter peak. Furthermore, the incidence of tropical storms is rapidly declining, having passed its September peak. The risk of running into very adverse weather over the sea is therefore probably at its minimum at this time. The main route of passerine migration is along the coastal 'flyway', taking place a full month earlier on a route which enables migrants to land in the event of poor weather – a strategy that oversea migrants cannot employ.

The tropical storms of the North Atlantic are known as hurricanes when they reach a certain intensity. Their tracks vary, some moving in a general westerly direction through the West Indies, while others turn north into the USA or along its east coast. A few curve northeast well out to sea. Landfall of a hurricane normally spells the beginning of its demise, since it relies on high sea surface temperatures for its sustenance. Those storms reaching higher latitudes off eastern North America die quickly over cooler seas, unless they engage polar air sweeping south, when they are occasionally re-invigorated into intense storms known as extra-tropical depressions (Fig. 46), which move across the North Atlantic in the upper westerlies.

The passage of a tropical storm across the path of migrants wreaks havoc, and reports abound of falls of exhausted birds on shipping. Mortality in a hurricane, and indeed other intense storms, must be very high over the sea. Looping trajectories in the fierce cyclonic winds with no chance of a landfall result in exhaustion and death. Some vessels have reported decks and rigging thickly covered with exhausted landbirds, and these include resident species blown out to sea in the very strong winds.

Waders

Wader migration is more difficult than most to correlate with the weather. Their flight – even that of the small species – is strong and fast, and some fly vast distances. Some arctic breeding waders in both Eurasia and North America reach as far as Australasia and Patagonia, travelling up to 13,000 km. In general, migration is begun in much the same optimum conditions as with other groups, fine anticyclonic weather in particular, and some appear to be less prone to drift by adverse winds than others.

Short distance sea crossings undertaken by Lapwings over the North Sea are made in favourable following winds, though, like other migrants, more passage in headwinds has been noted later in the autumn migration period, with such flights taking place at lower levels. Density of migration is greatest in following winds, decreasing steadily as winds change to strong headwinds when it is minimal.

In many of the long distance migrants, the date and route of migration may be determined only indirectly by environmental factors. This is illustrated by the fluctuations in the numbers of Temminck's Stints, Little Stints and Curlew Sandpipers in Britain. All are small north and east Palearctic waders, but numbers of the first named show no relationship to the others. Though much scarcer, it is more regular and may not be influenced as much by meteorological factors. In contrast, the Curlew Sandpiper shows marked fluctuations. 1969 was a year noted for remarkable passage in western Europe. Part of the reason for the influx, which peaked from mid August to mid September, was attributed to its then high level of breeding success. The weather in August, however, served to direct a large part of the population westwards from the normal migration route through the Baltic and across the Mediterranean. During the latter part of the month, a broad zone of low pressure from Scandinavia to the northern Urals, and high pressure over the Arctic Ocean, maintained periods of strong east to northeast winds over the regions bordering the Arctic Ocean and the Norwegian Sea. Slow-moving frontal systems south of this easterly flow maintained overcast weather. It was this persistent situation that coincided with the migration, and resulted in large numbers of Curlew Sandpipers being displaced west and southwest towards Britain. The two largest influxes corresponded with the two main periods of easterlies. The high proportion of juveniles suggest that, as in other groups, birds of the year are more prone to influence by the weather than adults.[164]

Invasions

Earlier, I mentioned briefly the fluctuations in autumn migration that arise due to non-meteorological variables, particularly population dynamics. There are certain species that are quite irregular in their movements, moving only in seasons of high population. Their abundance is normally due to successful breeding, perhaps in a fine summer following a mild winter with abundant food and low mortality. In these circumstances, overcrowding occurs, and when combined with a poor food supply after breeding has finished, considerable numbers, usually immature birds, erupt from the natal area and disperse widely. This allows a viable population of adult birds to remain and breed the next season. Variable numbers of this eruptive population may return in the following spring. Although this is a simplistic explanation for a rather complex phenomenon, it has been suggested[167] that eruptions take place every year and are stimulated by precisely the same meteorological factors as true migration. Unlike true migrants, however, movement may cease when the erupting element find food supplies, moving on again when these are exhausted. They are thus nomadic, and in years of abundant food move only short distances. The time of arrival at any particular place can therefore vary considerably, depending upon the size of the erupting population, the food supplies encountered during their wanderings, and the windflow.

The most frequent species to erupt are those with a specialised diet of food which itself fluctuates markedly in abundance – generally consisting of the seeds of perennial plants. Many of these birds come from the vast northern forests of Eurasia and North America, and feed on the seeds of conifers, birch and rowan. Unlike true migrants, they tend to move east-west to stay within the distribution of their food plants. Even some predators fluctuate in abundance according to the density of their prey; the Snowy Owl and Rough-legged Buzzard are two such species. There is also now considerable irrefutable evidence that large migrations of a wide variety of species (including true migrants) are also often dispersive in character when populations are high. This includes vagrants which travel over a long distance – in Britain for example, Nearctic and east Palearctic species (see Chapter 9). As we have already seen, dispersing birds tend to be non-oriented, but move downwind in fine anticyclonic weather. Some of the anticyclonic 'drift' movements described earlier are cases in point, where disorientation plays little or no part in determining the destination of birds. Whereas disorientation may be a significant cause of the grounding of true migrants, this cannot be the case in non-oriented birds. Landfall is merely at a site to which the windflow takes them, hence the occurrence of birds over a wide area after an oversea passage. It is often difficult to interpret some movements since a variety of factors may contribute to the passage. Both post-juvenile dispersal and reversed migration may occur alongside true migration, and be more widespread when populations are high.[187] Invasions are therefore not necessarily related to unusual weather conditions, but occur when anticyclonic weather coincides with the need to seek food resources away from the natal area. Several invasion species from one region often appear in numbers together – all having been influenced by fine weather during and at the close of the breeding season.

Of the recognised invasion species, two stand out, the Crossbill and Waxwing. Both are stimulated to erupt in the conditions described above, and large arrivals in Britain have invariably taken place in light winds from various directions – chiefly those with an easterly component. Even when the seed crop is good, small eruptions may occur when it is finished, so that the size of the crop not only has a role in determining whether there will be an eruption, but also its magnitude and timing. During these movements, the origin of Crossbills (as with some other species) can often be determined by examining the morphological characters of trapped birds or corpses. Several species vary in a small but definite way over their breeding ranges, and the movement of these various forms is controlled by the position and intensity of high pressure systems. Other northern forest birds irrupt into Britain – as varied as the Bullfinch, Great Spotted Woodpecker and Nutcracker. The last is very rare, but in 1968, in response to a scarcity of its staple diet of Arolla Pine seeds, the thin-billed form of northwest Russia and western Siberia erupted in great numbers over wide areas of Eurasia, particularly to the Baltic, Germany and the Netherlands. On the Swedish coast in one three-hour period, 4,500 birds flew west under the influence of the axis of the Scandinavian anticyclone. During anticyclonic spells in August and September, large and unprecedented influxes took place into Britain along the east coast between Kent and the Wash. Ringing recoveries have shown that even Waxwings irrupt into northern Europe from as far away as Siberia.

Less familiar species in an eruptive context are the tits, the Fieldfare, some finches and the Bearded Tit. Whereas eruptions of forest birds are rarely seen in the earliest stages, the Bearded Tit is one bird which is easily observed, since eruptions begin with excited birds rising, calling, from reedbeds. This behaviour is noticeable chiefly in light winds of force 2 or less (less than 4 m/s) in sunny anticyclonic weather. Eruptions also occur in overcast weather with poor visibility, but nonetheless, are still associated with light winds. Eruptions are delayed or reduced when late autumn weather is dominated by mobile depressions. In the anticyclonic October of 1972, when a blocking high lay over, or to the west of, northern Britain, several hundred birds were observed in some southern English reedbeds, and small flocks even reached the Scillies and southwest Ireland. Such wandering birds are very restless, and move great distances downwind. The Dunnock is one of the many relatively resident species in Britain to show eruptive behaviour, usually over short distances and loth to cross the sea. Migratory activity is shown under similar circumstances as in the Bearded Tit, in winds of not more than force 3 (5 m/s).

In January 1937, an immigration of Fieldfares in Greenland (see Chapter 11) followed a hard-weather movement from northern Europe. Svardson[167] thought that the time lag between arrivals in northeast Greenland and Jan Mayen, and arrivals at the southern tip of Greenland (bearing in mind the date) indicated two different events, and not a southward movement within Greenland, since the northern birds would have had difficulty in surviving for long enough. That season was noted for large eruptions of Fieldfares over northern Eurasia, and he maintained that it was more likely for the northernmost birds to have been part of a mass eruption west and southwest from northern Russia.

Finally, one species which erupts under rather different conditions to those described above is the Pallas's Sandgrouse. It is rather a historical invader, since its breeding range in the arid regions of southwest Asia has withdrawn in the past few decades. Eruptions of this bird are a necessity, involving large numbers whose grass seed diet becomes unavailable. This may be due to drought, or unusually deep or hard-crusted snow. Movements due to the latter are therefore of the nature of long distance hard weather movements (see Chapter 11), the last irruption to western Europe – albeit small – being in 1969. Invasions of northern grouse may also take place when their food supply is covered by very deep snow.

DISPLACEMENT IN SPRING

The return of migrants in spring has an urgency that is lacking in autumn – at least in early autumn – but they are approaching regions where the weather is improving progressively. Summer visitors to Europe which winter south of the Sahara overfly the desert as in autumn. Many move along the western fringe where conditions are slightly better, particularly in the north where the rainy season is coming to an end.

Over northwest Africa, there is a high probability of adverse weather in early spring associated with depressions on southern tracks. Even on the desert fringe itself, poor weather contributes towards the late arrival of European summer visitors by causing delays and sometimes heavy mortality. There can be considerable drift over the desert, and easterly winds carry migrants into the Atlantic. April can be very cold in Morocco, and in 1974, I saw very weak hirundines being killed by vehicles when flying low in persistent cold wet weather. Such conditions give snow over the Atlas mountain barrier, and reversed passage may even take place. The 1974 event, caused by outbreaks of polar air behind depressions moving across northern Morocco, was followed by a noticeably late arrival in Britain of Swifts, hirundines and one or two other small insectivores. Extensive cloud on the associated cold fronts gave heavy rain over coastal Morocco and snow over high ground. Following delayed migration, sudden upsurges in numbers often occur when migrants finally arrive at their northern breeding grounds. Reversed migration in spring in falling temperatures improves survival, but drift and grounding by adverse winds, cold weather and precipitation all take place as in autumn.

A blocking anticyclone to the west of northwest Europe brings cold springs and delays northward migration. The winds on its eastern flank are from the north, often bringing disturbances south from the Iceland and Norwegian Sea areas. Such a block persisted for two weeks in mid April 1973, centred between 10°W and 30°W (Fig. 29). The block collapsed at the end of the month, with a filling depression moving into Biscay finally bringing warm south to southwest winds into the Iberian peninsula and allowing a sudden surge of migrants into northwest Europe in the last few days of April. David Lack[84] found that weather influenced Swifts little on the last lap of their northward journey, except that major delays occurred when cold northerlies prevailed during their main arrival period. Migrants meeting adverse weather near their destination often suddenly cease passage, and even those which have reached it may make reversed movements. During a cold spell, for

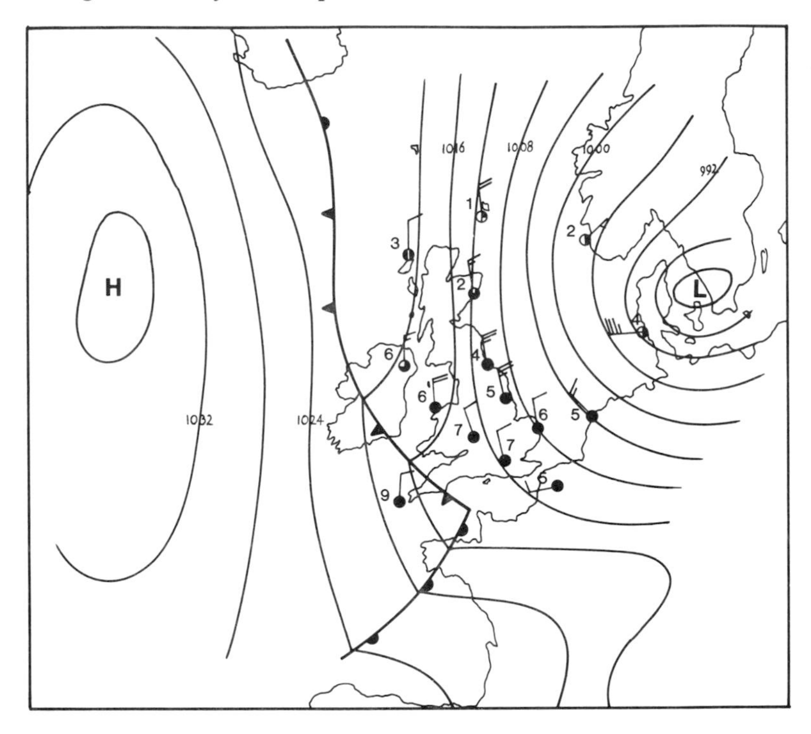

Fig. 29 Typical situation delaying spring migration 11th April, 1973, 0600 hours. Blocking anticyclone west of Britain with cold northerlies prevailing over Britain and the North Sea. Frontal troughs running south in flow. Temperatures in °C.

example, Sand Martins may move southwest from southern England towards the milder southwest peninsula. Many large falls along the channel coast of Britain owe their substance to southward moving cold fronts, and even departing Scandinavian winter visitors are affected. Similar falls occur in North America as northbound migrants encounter cold fronts when crossing the northern shores of the Gulf of Mexico. A spell of very cold weather immediately succeeding the arrival of summer visitors can lead to serious mortality. In Britain this is rare, but in the more continental climates of North America and Eurasia, it is of more regular occurrence for the earlier small migrants bound for high latitudes and montane regions.

Equally, spring migrants can arrive very early. In early March 1977, a host of summer visitors appeared in southern England. Besides the usual early birds – Wheatears, Chiffchaffs and Sand Martins – there were Cuckoos, Swallows and Hoopoes. A 'high' over the western Mediterranean from late February to 8th March gave warm southerlies and high temperatures, and the origin of the airmass was indicated by falls of Saharan dust in Northern Ireland and western Scotland.

There is a tendency for spring migrants to move in poorer conditions than in autumn, and passerines cross the North Sea more often against the wind.[85] Drift can therefore be more extensive, and falls of small night migrants in northern Britain (especially the islands) are common. Since these migrants chiefly comprise Scandinavian summer visitors, the oversea drift becomes progressively more pronounced with latitude. Populations do not include inexperienced juveniles setting out on their first journey, however, and falls of migrants are smaller than in autumn. Blackcaps and Whitethroats in northeast Britain are more likely to be Scandinavian-bound birds than those arriving further south and west, where most are British birds. Species which are scarce breeders in Britain thus occur more frequently in the north, such as the Pied Flycatcher, which is a much more common spring migrant on Fair Isle than in eastern England.

A substantial fall of spring migrants appeared on Fair Isle on 3rd May,

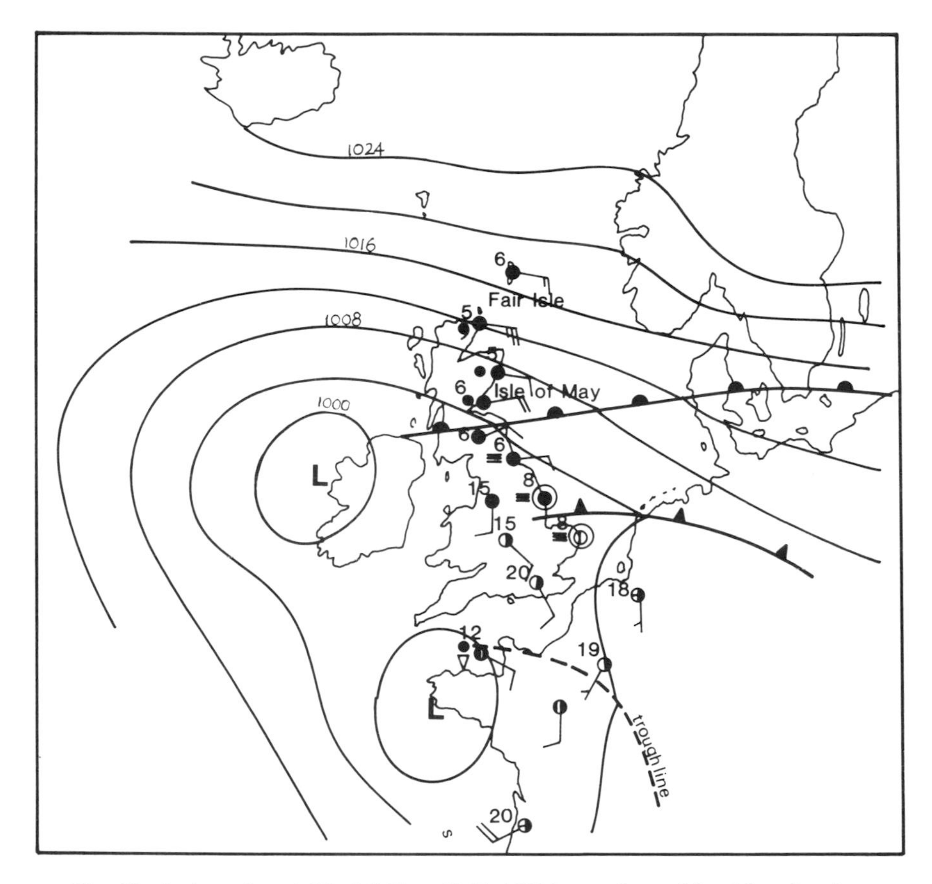

Fig. 30 Spring migrant fall, 3rd May, 1969, 1200 hours. Large falls of Scandinavian-bound spring migrants in northern Scotland.

1969.[32] Southerly winds over Iberia and France on 1st and 2nd May had pushed warm air into the southern North Sea. The warm front moved slowly north to lie from southern Scotland to Denmark by midday on 3rd May (Fig. 30). Strong easterlies prevailed ahead of the front, and migrants flying north overtook the front to penetrate the cloudy, misty and wet weather over the northern North Sea. When the precipitation at Fair Isle died out at midday, one of the most spectacular spring falls on record began. Among the birds making landfall that day were 300 Ring Ouzels, 45 Wrynecks, 400 Willow Warblers, at least 1,500 Tree Pipits, 500 Bramblings and 32 Ortolan Buntings. These were adults – the males resplendent in full breeding plumage – in an event for which this small island is famed. The front remained to the south and weakened, so that by 6th May skies were clearing. Winds had decreased considerably, and that night there was an exodus as many birds departed for their Scandinavian breeding grounds. The fall was echoed further south on the Isle of May, which experienced extensive sea fog that month – a frequent phenomenon accompanying such falls, since it is brought by warm moist onshore winds. Surprisingly, a large Tree Pipit arrival occurred a day later than that on Fair Isle – an indication of the ongoing passage and slight changes in the meteorological situation.

A year later, almost to the day, a similar fall took place – but this time it was more extensive along the east coast of Britain, with numbers increasing northwards where the winds were stronger. This more widespread distribution was to be expected, since northeast-bound migrants experienced progressively greater drift in backing southeast winds. These winds lay on the southwestern flank of a persistent anticyclone over Scandinavia, freshening steadily as a weak depression in Biscay deepened and moved into West Germany by the 8th May. The easterlies were warm and humid, with extensive sea fog on eastern coasts, but fine weather in the extreme north suggested drift in strong crosswinds, rather than disorientation. Falls of Scandinavian migrants continued for several days, mainly between 4th and 13th May. Not a typical disorientation situation, although in the south the weather was certainly much poorer, with outbreaks of heavy rain and thunderstorms over eastern Scotland. A feature of this fall was the large number of Bluethroats of the red-spotted form – the form that breeds in Scandinavia north of 60°N, and another indication of the intended destination of the migrants.

Long oversea journeys in spring pose greater problems than in autumn, since the optimum conditions may not be as persistent. As shown in Chapter 7, suitable tail winds are found on the western flank of high pressure zones, but this is an area prone to deterioration of weather from the west. Swans and geese leaving northern Scotland for Iceland and Greenland are naturally unable to predict conditions over their intended route.[189] The optimum situation of a blocking high over the route is prominent between longitudes 20°W and the Greenwich meridian from March to May, though if the centre of the anticyclone is at a more southerly point, the risk of disturbances moving round the northern flank is high. Birds departing under these circumstances may find it necessary to turn back, but in the southeasterlies ahead of an approaching depression, this only succeeds if they are close enough to land. Problems may also arise if they run into winds from an adverse direction e.g. if a depression passes to the north.

As in autumn, the drift that some migrants undergo can be very marked. One June, a British-ringed Swallow was recovered 320 km SSW of Ireland in strong pre-frontal southeasterlies. Two May birds, however, were ringed in Ireland after a strong southerly wind, and recovered in France and east Germany within two weeks, having corrected for their displacement. Birds of other species have been carried westwards across the Atlantic on ships after displacement by easterly winds, and in April 1978, a Scottish-ringed Osprey was found on a boat in the Denmark Strait off Greenland, after prolonged southeast to east winds over the northeast Atlantic.

One aspect of migration peculiar to spring is the phenomenon of overshooting. This is undertaken by a wide range of species, some of which re-orient (giving the impression of reversed passage), but it is more noticeable and typical in Britain of southern birds which overfly their destination in unusually fine anticyclonic weather, particularly if temperatures are well above normal. The areas in which overshooting birds appear are dependent on their orientation. Those from the south are more often found in southern Britain, while those from the southeast arrive in east coast areas, sometimes as far north as Shetland.

Of southern species, Garganey, Hoopoe, Golden Oriole and Quail are well known periodical migrants in warm anticyclonic conditions. Quail occasionally reach invasion proportions – 1953, 1964 and 1970 were years of particular abundance. In the last mentioned year, May and June were abnormally anticyclonic over the whole of Europe, especially in the latter month when temperatures in southern Britain were as much as 2°C higher than normal. Many Quail remained to breed.

OVERWINTERING

Early arrivals of individual migrants are occasionally difficult to separate from those overwintering. One by-product of mild winters is the wintering of birds of typical long-distance migrant species, breeding in northern Europe but normally wintering in Africa. The winter of 1974/75 was one of the mildest for decades, with December and January being dominated by an Icelandic low and Azores high both more intense and further northeast than normal. In the very strong WSW airflow over Britain, warm sectors were frequent, and temperatures rose some 3 to 4°C above average. In January, apart from the unusual sight of Whinchats, Swallows, House Martins, and single Turtle Dove, Cuckoo, Redstart, Garden Warbler, Whitethroat and even Yellow-browed Warbler, birdwatchers saw an adult Black Tern and a flock of Common or Arctic Terns. All these were mostly in southern England. One of the Garden Warblers was an off-passage migrant ringed at Spurn Head in the previous October. Blackcaps and Chiffchaffs were more widespread than usual throughout the British Isles, and ringing shows that these are generally of Continental rather than British origin. Numbers of wintering Blackcaps have been directly correlated with the abundance of autumn passage migrants on the east coast, and also the improved survival brought about by the recent adaptation to taking foods provided by man.[90]

SUMMARY

In this chapter I have described the weather-related movements that migrant landbirds, waterfowl and waders make away from their normal routes. Their propensity for displacement varies with species, age and experience, the degree of orientation and their diurnal migratory rhythm. Disorientation occurs when navigation is impaired – deep extensive cloud systems, precipitation and poor visibility associated with fronts are three contributory factors. If disorientated for any length of time, downwind flight ensues and, if landfall cannot be made, drift is marked and exhaustion and mortality may result. Drift, without disorientation, may affect any migrant if winds are strong enough. Onward passage, or perhaps re-orientation, is later achieved in favourable conditions. A landfall of disoriented or drifted migrants may result in large numbers of birds arriving in localised areas.

More widespread falls of autumn migrants take place in less adverse weather, usually associated with downwind flight in anticyclonic situations. Since such conditions are not conducive to disorientation, these movements are thought to involve largely non-oriented birds – probably juveniles on post-fledging dispersal. Reverse orientation is also shown by certain species in response to unseasonably fine and warm weather.

Falls of migrants may include birds from one or several different types of movement, making interpretation difficult on some occasions. Population fluctuations affect numbers of migrants, and in years of abundance, post-juvenile dispersal is widespread. Species with specialised diets and which do not show regular migrations, also disperse when populations are high. These eruptions are stimulated by anticyclonic weather, and direction of movement is guided by the wind.

Similar falls due to disorientation or drift take place in spring. Severe weather causes reversed movements and delays arrivals at breeding grounds. Conversely, fine warm weather may cause overshooting of certain species beyond their breeding range in their standard direction. An abnormally mild winter induces some off-passage migrants to remain for the season.

CHAPTER NINE

Vagrancy

Arguably one of the more exciting events of a birdwatcher's career is the discovery of an individual bird which is clearly thousands of kilometres from the edge of its range or migration route. Such birds, known as 'extra-limital' species, or vagrants, are of little scientific interest on their own, but careful watching, recording, ageing and sexing of each bird, followed by an analysis of the details of all the records, can result in important inferences. What appear to be random occurrences may fall into a pattern which gives a clue to the movements and population dynamics of a particular species.

There is, in fact, evidence to suggest that the records of at least some vagrants are linked to population changes, but the eventual appearance of an individual, long distances from its normal range, is invariably correlated with the weather. Attempts have been made to correlate variation in the abundance of vagrants in Britain with climatic change. While this may be possible in some instances, the fluctuations described in this chapter are due primarily to climatic variability, and caution must be used when looking at long term trends.

It is often difficult to link the appearance of vagrants to the meteorological situation over their assumed routes. This is particularly so when their route lies over terrain which affords a landing place at any time during their flight, since one cannot know how long the flight has been in progress, and to what extent a bird has rested en route. Perhaps the best way to analyse records of land birds approaching from a landward direction is to compare seasons of comparative abundance and scarcity with variations in prevailing weather patterns.

On the other hand, the study of the arrival of land birds which are known to have made transoceanic crossings is facilitated by the knowledge that in most instances the bird could not have landed on the sea. Apart from a few ship-borne vagrants, their flights must therefore have been continuous. A broad knowledge of their altitude preferences is also of assistance, although one cannot make too many assumptions.

Vagrancy is, of course, a relative term. An abundant breeding bird in one region may be only a vagrant in another, while short distance migrants or sedentary species may be vagrants only a few hundred kilometres from their normal range. While vagrancy is most prevalent during migration seasons, some individuals or species wander at other seasons, especially during winter hard weather movements (Chapter 11).

PALEARCTIC VAGRANTS

Of the untold millions of migrant passerines from Central Asia and Siberia that winter in southern Asia, a few individuals find their way each year to NW Europe and Britain. Falls of over 30 individuals were noted in 1975, 1981, 1982 and 1985, and have included Pallas's, Radde's and Dusky Warblers, as well as thrushes and buntings. Their appearance has been correlated with the distribution and intensity of the high pressure zone which begins to form over Siberia in early autumn.[7] (Fig. 50). This anticyclone waxes and wanes, and when most prominent creates an extensive easterly flow on its southern flank. Whether on reversed passage, or merely post-juvenile dispersal stimulated by the fine weather and sustained by the easterlies, is in doubt, but nonetheless, passage of immature birds takes place westwards with further movement across Russia and northern Europe dependent on synoptic situations similar to those influencing European migrants.

Correlation of influxes of far-eastern vagrants into Britain with westward extensions of the Siberian high is greatest when combined with depressions moving into Europe on tracks south of 55°N. (e.g. Fig. 28). These steer such vagrants around their northern flanks, contributing towards a markedly more northern distribution of records in northern Europe than in Britain, although the 1985 fall appeared to be due to continued westward passage in the easterlies of a persistent anticyclone over Britain and central Europe.

Some vagrants from the nearer regions of northeast and east Europe appear to exhibit reversed passage in warm wind regimes opposed to their normal E to SE heading, notably the Scarlet Rosefinch, Greenish and Arctic Warblers. These wind flows carry them via Scandinavia on the northern flanks of depressions.

Other warblers from more southern areas in Europe appear in fine weather in autumn. These include Subalpine and Aquatic Warblers. One theory suggests that individuals of the latter species are reversed passage migrants from the Italian breeding population, but the weather during the peak year of 1972, when there was an abnormally high degree of anticyclonic weather from July to September, indicated a post-juvenile dispersal from Baltic breeding grounds. With high pressure prevailing over northern Britain, periods of fresh east to northeast winds occurred in all three months.

The return of east European birds in spring, northwestwards from Asian wintering grounds, may result in overshooting during fine weather. In the large falls of Scandinavian migrants of May 1970, described in the previous chapter, the appearance of several rare eastern passerines suggested that overshooting had indeed taken place in the warm easterlies. Among these were Thrush Nightingales, Greenish and Dusky Warblers, and Scarlet Rosefinches, as well as lesser rarities such as Icterine and Barred Warblers and Red-breasted Flycatchers. Since their standard migratory direction in spring is west to northwest, it is not difficult, given the right conditions, for birds to overshoot their breeding range, even though some are extremely rare at this season. A remarkable example of overshooting on spring migration, and subsequent reorientation over a long distance, concerned a Rustic Bunting ringed on Fair Isle in mid June 1963 and recovered the following October on one of the Greek islands. This species, with only a few records in Britain, normally migrates just north of west from its Asiatic wintering grounds to the breeding area in the Baltic states.

Another group prone to overshooting in spring (besides some terns, to be discussed in Chapter 12) is the heron family. The Little Egret and Purple Heron are the most frequent species, followed by Night Heron and Little Bittern. Again, the stimulus is fine weather over their destination, and in 1970, a spring noted for its anticyclonicity, three days of high pressure over Iberia in mid April resulted in widespread arrivals of these species, mainly in southwest England. Their route appears to have taken them over the southwest approaches, and other southern migrants, Hoopoes and Alpine Swifts, accompanied them.

Such a mixture of southern vagrants in mid-May 1979 coincided with a deep warm southerly airmass which deposited Saharan desert dust in N. Ireland. These dust-bearing airmasses are rare and irregular events, and show little seasonality. They do display their origin quite clearly and can be linked with vagrancy from the south on certain occasions, even of rare moths. Another notable dust fall in mid-November 1984 coincided with the arrival of several Pallid Swifts and a Desert Wheatear, extreme rarities of which the latter is at best a partial migrant breeding in subsaharan North Africa. This late date suggests a remarkable reverse movement for the swifts, most of which by then should be wintering in tropical Africa.[201]

A similar weather pattern has been linked to the few arrivals this century of the only tropical African bird to reach northwest Europe – the Allen's Gallinule. It has been observed only a handful of times, but its appearance in winter has been correlated with disturbed weather succeeding an anticyclonic spell with southerly winds blowing from North Africa. In common with other birds of its region, breeding is controlled by seasonal rains. In the northern tropics, an apparent dispersal takes place at the beginning of the dry season in December, in keeping with the few European records.

These birds share with others of the family a habit of wandering. The Corncrake has been recorded about twenty times from Greenland, blown off course by storms, while the American Purple Gallinule has been found in many regions bordering both North and South Atlantic.

NEARCTIC VAGRANTS

Transatlantic vagrancy is a phenomenon remarkable not only for the distances covered, but also for the fact that it involves an oceanic crossing of up to 5,000 km. In previous chapters I described the considerable late autumn migration on an oversea route from northeastern parts of North America to the West Indies and South America, in which several land-bird species make use of strong following winds. Disorientation and displacement within developing frontal waves off the eastern seaboard are mechanisms which occasionally result in transatlantic crossings.

In a recent study,[44] I was able to correlate arrivals of Nearctic landbirds with certain meteorological situations. Falls, albeit small, of such birds in northwest Europe occur only when wave depressions move rapidly across the Atlantic without developing sufficiently to become occluded and slow down. The formation of these was described in the previous chapter. Migrants disorientated over the western Atlantic are drifted by the strong southwest winds to the south of such a wave depression. The wave depression theory is contrary to some past opinions which maintained that hurricanes or ship-assistance were the chief causes of vagrancy.[38] Hurricanes and deep slow moving depressions are each more likely to carry displaced birds into looping trajectories round the centres, in which, in the absence of a landfall, they perish. It is nevertheless possible that the few vagrants appearing in Iceland are carried there in such depressions. Naturally, the landfall of any individual depends upon its position in relation to landmasses and the wind system in which it flies. Those reaching Britain remain in the strong, often broad, southwesterly warm sectors which run rapidly (within one to three days) across the Atlantic. Between 1967 and 1976, over 70% of the Nearctic landbird vagrants in Britain and Ireland could be associated with such airflows. There have even been instances of migrant American moths and butterflies crossing the Atlantic under similar circumstances.

This particular type of weather system takes a southern track across the ocean, and calculations indicate that the majority of vagrants begin their crossing due to displacement well south of latitude 45°N, following a trajectory that only reaches 50°N close to Britain. This accounts for the

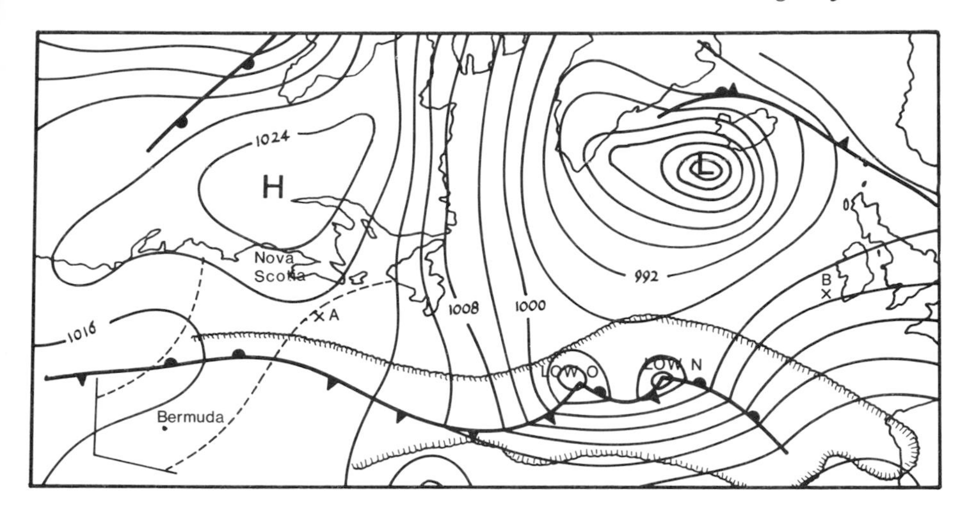

Fig. 31 Transatlantic landbird vagrancy to Britain 4th October, 1976, 1200 hours. A shows position of low O at 1200 hours on 3rd October; B shows position of low N at 0600 hours on 5th October; hatched line encloses area of deep frontal cloud associated with lows O and N. Broad arrow shows normal route of oversea migrants (from Elkins 1979b)[44].

relative abundance of records in extreme southwest Britain and Ireland, where they probably arrive at a low altitude because of exhaustion. The dearth of vagrants in the first half of September, even in apparently 'ideal' weather conditions, is related to the timing of the oversea migration, which peaks much later than that overland.[145]

One of the most notable autumns for such vagrancy was that of 1976, when 25 individuals were recorded. Over 60% of these were northern species displaced while on normal southward migration over the sea – Grey-cheeked Thrush, Yellow-rumped Warbler and Blackpoll Warbler. This large unprecedented fall (there were 14 of the last mentioned species, of which most appeared in multiple arrivals in the first nine days of October) was attributable to the abnormal atmospheric circulation. A proximate cause of the long British drought which ended in September of that year was the extreme northern position of the polar front jet stream. The jet suddenly shifted south from this position in September, and, in combination with other factors resulting from the long hot summer, this produced unusually vigorous cyclonic activity which was ideally suited to transatlantic vagrancy in October.

Figure 31 shows the anticyclone over eastern Canada affording fine migration weather over that area. A waving front across the migration route generated two small depressions which initiated a strong westerly airflow in their broad warm sector. The rapid movement of the system was responsible for the arrival of at least six Blackpoll Warblers in southwest Britain between the 6th and 9th October. Further fast moving warm sectors brought more vagrants later in the month.

Other landbird vagrants are thought to encounter such frontal waves while on reversed passage. Reversed northeastwards passage of migrants in eastern North America has been noted in warm airstreams from a southerly point, and over Nova Scotia has been seen on radar even in winds from between south and west behind a cold front.[145] Some of the Nearctic vagrant landbirds depart in subtropical airmasses south of a cold front, only to be caught up in an intensifying wave. Unlike the northern species on southward migration over the sea, these have a more southern distribution and are rather earlier, including such birds as the Red-eyed Vireo and Northern Oriole. This was the mechanism of transatlantic crossing suggested in an earlier study by Ian Nisbet,[124] and supported by events in 1967 and 1975, when the species composition of vagrants was materially different to that in 1976. In 1975, warm southerlies in eastern USA, thrown up by the dying hurricane 'Eloise' inland, met polar air to the north and gave extensive cloud and rain over 12 degrees of latitude. Reversed migration from as far south as 35°N was thought to have taken birds into the developing westerlies south of this waving front. Such frontal situations have coincided with lulls in the normal southward passage of coastal migrants, and this event was no exception. A huge movement of pent-up migrants was evident as the front finally cleared the area.

Lest it be thought that all North American landbird vagrants in Europe cross unaided, it should be pointed out that there are at least a few crossings which can be attributed to ship assistance, the birds having landed on eastbound vessels while migrating through tropical and other storms. Alan Durand, who travelled frequently on ocean liners plying between England and the USA, listed very many instances of shipboard waifs.[37,38] Those landing on eastbound ships sometimes survive a transatlantic crossing, particularly those able to be fed, though even insectivores exist on migrant moths and butterflies. A vessel sailing with the wind benefits the migrants by reducing the airflow over the decks, enabling them to settle. Most have no chance of doing so if the ship is steaming into a strong wind.

Although the absence of oversea migration off the eastern seaboard of the USA in spring leads to little risk of an inadvertant transatlantic crossing, there are occasions when coastal migrants overshoot or meet frontal weather in which strong winds drift them out to sea. A few of the spring vagrants in Britain can be traced to such situations.[44] It is not as clear a correlation as in autumn, with a puzzling dominance of sparrows and buntings among the vagrants. Their appearance in Britain has generally coincided with anticyclonic weather, and sightings much further north than in autumn suggest an onward passage after an initial undetected arrival. The only 'fall' that has been associated with fast eastward-moving waves was that in May 1977, when four individuals were recorded in Britain. An active waving cold front had crossed the southeast USA about a week prior to the first record, with the polar air giving record low May temperatures over the Carolinas. The strong warm airflow south of the front extended fairly rapidly to Britain, and it is possible that many northbound migrants reaching northern Florida were grounded or drifted out to sea.

The only Nearctic landbird that has appeared more than a handful of times in Britain in winter is the American Robin, of which there is a large gregarious population wintering in southeastern USA. Any movement, such as might take place in cold weather, may risk a flight out to sea in an area where winter formation of depressions is high along the Gulf Stream boundary (Chapter 8). Transatlantic crossings in warm sectors are then possible.[44]

The species composition of the not infrequent falls of Nearctic waders in western Europe throws up some puzzling questions. Arrivals in Britain reach a peak a full month sooner than the Nearctic landbirds, and their appearance is not directly related to the wave depressions mentioned earlier. Additionally, some of the most abundant species are not the oversea migrants, but those which migrate chiefly through the interior of the North American continent and are thus relatively scarce on its eastern seaboard. Some of the arctic waders may overshoot and be carried across the Atlantic in strong winds, particularly if they are flying within the jet streams over eastern Canada (Chapter 7), but this is purely speculative and difficult to substantiate. Such birds would include the Pectoral Sandpiper, Lesser Yellowlegs and Lesser Golden Plover. Vagrants from northeast Siberia (the Sharp-tailed Sandpiper, for example) are regular in Alaska, and may also arrive in Europe via the route taken by Canadian arctic species. Waders over Nova Scotia fly higher than landbirds, generally between 1 km and 3·5 km, even up to 6 or 7 km, thus subjecting them to the strong upper winds. Differences in species composition in Britain may indeed reflect variations in flight altitude, though year to year changes will also reflect population fluctuations and the timing of migration in relation to jet stream positions and strength.

The inland migratory waders rarely reach the Atlantic seaboard in large numbers, and some are almost as common in Britain. Falls of these have coincided with falls in eastern USA and Canada, and one meteorological situation apparently common to most large arrivals in Britain (especially of Buff-breasted Sandpipers and Wilson's Phalaropes) is the passage of active weather systems from central USA to the Great Lakes and Labrador beneath a strong southwesterly upper windflow. In the peak years of 1975, 1977 and 1980, the polar front jet in September was stronger than normal or displaced further south. Although again only speculative, I think it possible that southbound migrant waders climbing to the rear of these disturbances are carried northeast and east by the upper winds, some moving only as far as the Atlantic coast, but others making a transatlantic crossing. This is particularly valid for the Wilson's Phalarope, which has a central breeding distribution in North America. In 1976, when the polar jet was abnormally displaced, and the Great Lakes region was dominated by anticyclonic weather in August, very few of these inland migrants appeared in Britain.

It would seem, therefore, that it is the upper transatlantic windflow that carries Nearctic waders (and probably also wildfowl) to Europe. Inland migrants are perhaps less able to compensate for drift than oceanic migrants, although it is possible that most vagrants to Europe are poorly-oriented birds, e.g. juveniles with faulty or poorly-developed orientation systems.

There are, of course, numerous species other than those discussed that are carried across the Atlantic, in both directions and at most seasons. While these records are of interest, they are of little significance, except perhaps for one reason. Range expansion has occurred in this manner, for Spotted Sandpipers from North America have bred in Scotland, and Black Duck have hybridised with Mallard in England. There is also the fascinating possibility of other Nearctic birds now migrating within Europe after a transatlantic crossing, so that sporadic breeding may occur again. Overwintering of land-bird vagrants has now taken place, and winter and spring observations of Nearctic waders have become more frequent. However, one species, the Killdeer, is subject to displacement by winter storms in eastern USA,[26] resulting in an arrival pattern reminiscent of that of the American Robin, and cold weather movements may also have been the origin of recent winter influxes of Ring-billed Gulls. Range expansion has also taken place in the other direction, with Cattle Egrets of European or African origin now breeding in the Americas. This species may have crossed with assistance from the constant winds of the Trade wind zone, and in this context, it is of interest that a greater number of Palearctic species, especially waders, have made landfall on the island of Barbados than on the eastern seaboard of the USA.

SUMMARY

Individual birds are very occasionally found great distances from their normal range, causing great excitement among ornithologists. Their arrival is normally the result of weather situations which act either on dispersing or wandering birds with perhaps poor navigational ability, or in an extreme manner on migrants on normal passage. Annual fluctuations may also be linked to population changes.

CHAPTER TEN

Migration of soaring birds

Deserving of a short chapter of its own is the migration of soaring birds. In Chapter 2, I described static soaring, the physical attributes of soaring birds, and the meteorological aspects of upcurrents. The formation of such airflows and their distribution clearly have an important influence on soaring migrants, and the following should be read in conjunction with the earlier chapter. Many migratory broad-winged species use upcurrents to minimise energy expenditure while on long distance cross-country flight. They use well-defined and often quite narrow routes, taking advantage of unchanging topographical features which provide sustained orographic uplift and reliable thermal sources. This precludes the use of long sea crossings (Fig. 32). There are some species less dependent upon upcurrents than others, but in general, the birds in this category comprise storks, cranes and a large variety of birds of prey.

STORKS AND CRANES

The two European storks are both migratory. The White Stork is mostly diurnal in its migratory habits, since it uses thermals extensively, but the Black Stork has narrower wings and is therefore less dependent upon

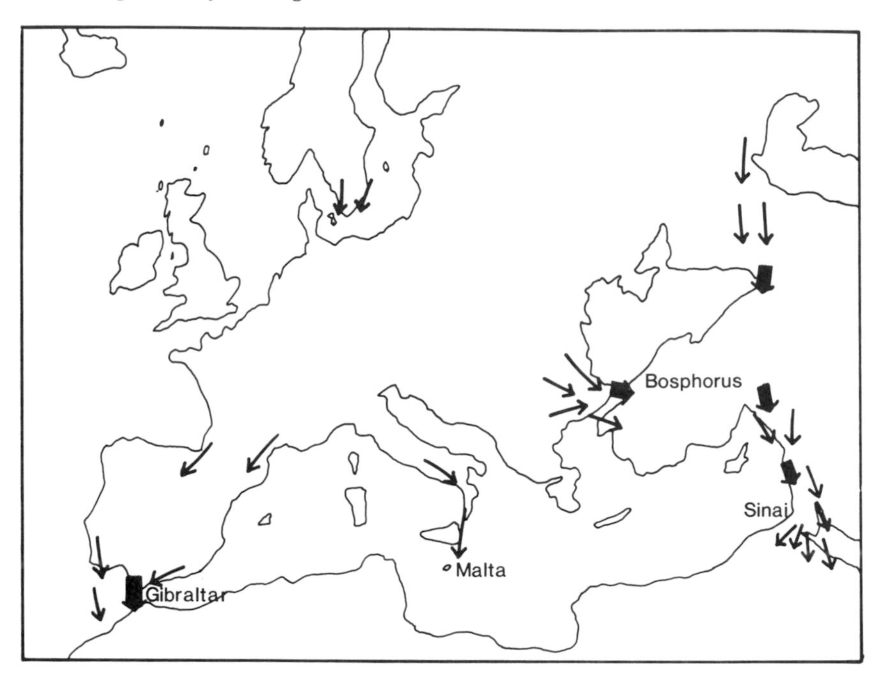

Fig. 32 Major migration routes for soaring birds into Africa (from Newton 1979)[122].

thermals. Nevertheless, although the latter can rise to considerable heights by flapping, and will cross wider stretches of water, thermal soaring remains its favourite mode of travel. The White Stork uses traditional routes crossing the Strait of Gibraltar and the Bosphorus, at each end of the Mediterranean Sea. Both allow entry into North Africa and a crossing of arid regions with their abundance of dry thermals. The Bosphorus route channels a huge migration round the eastern end of the Mediterranean along a narrow corridor and into Egypt. Dust storms occasionally cause considerable mortality, particularly the strong dust-laden southerly wind known as the khamsin[107] – most frequent in spring ahead of eastward-moving Mediterranean depressions. Although chiefly diurnal, nocturnal migration has been recorded in Algeria in the intense upcurrents produced by the flares of oil stacks.

One of the distinguishing features that separates migrating storks from cranes at long range is the latter's direct flight, often in V-formation, and the European Crane is certainly able to travel independently of upcurrents. Nevertheless, when thermals are available, migrating flocks combine direct flight with soaring, and are able to fly over land and sea with equal efficiency. As in other birds, migration is inhibited by opposing winds and rain, and northbound Cranes in spring are delayed by low temperatures. Their groundspeed on migration is erratic, according to the method of flight, and in thermals, the horizontal distance covered in a given time is much less than when in direct flight. In spring, Cranes over southern Sweden appear to compensate very successfully for drift overland by using landmarks, but over the sea they are more influenced by crosswinds. Drift in thermals can be quite considerable, since the birds must remain with the rising air in order to climb, even though it may be moving in a direction not in accordance with their heading.[2] The degree of drift when landmarks are invisible is illustrated by the remarkable influx of Cranes into southern Britain from 29th October, 1963, which was also recorded in West Germany and the Low countries. 500 or more birds were observed in Britain, having been drifted by southeasterlies strengthening as an anticyclone retreated eastwards over the Baltic Sea. Widespread mist and fog covering northwest Europe, together with an area of thick cloud aloft, may have prevented them from compensating for the westwards drift. Onward passage was undertaken when a showery airmass with broken cloud became established in early November.

RAPTORS

By far the greatest number of soaring migrants consists of broad-winged raptors, particularly buzzards, eagles, kites and hawks. While many raptors remain in Europe during winter, those that feed on cold-blooded prey – insects, reptiles and fresh-water fish (unavailable in frozen lakes) – must leave northern temperate and polar regions completely. These birds tend to migrate early in autumn, and like other migrants, their departure is stimulated by clear skies and following winds.[122] The broad-winged birds migrate mostly in the middle of the day, particularly when a dearth of orographic uplift over their route forces them to be totally dependent on the existence of thermals. The commencement of coastal passage has, however, been correlated with the onset of the sea breeze, with its attendant convergence zone. When thermals are restricted, the use of standing waves necessitates passage at lower altitudes. Maximum visible numbers then occur, just as songbird passage appears heavier when poor weather compels migrants to descend.[208] Some migrant raptors do not require upcurrents, and even among the broad-winged species, some are less dependent than others. This enables them to cross larger stretches of water, although a tendency remains to use narrow straits. Harriers and falcons use thermals very little on migration, especially the more powerful fliers such as the Peregrine. Harriers have been recorded flapping across the Sahara desert only about 15 m above the ground.

The Honey Buzzard is an abundant long-distance migrant that uses longer sea crossings than one might expect. Although exploiting upcurrents freely, it is capable of sustained flapping flight against strong opposing winds over wide stretches of water. Its independence of upcurrents is especially illustrated by studies of migration over Malta, where, accompanied by falcons and harriers, it is the only frequent broad-winged raptor. The sea crossings here are much wider than at Gibraltar and the Bosphorus, and there is a high correlation between raptor passage and the light winds, clear skies and stability of anticyclonic weather. The persistence of this weather type prior to mid October enables successful crossings to be completed, and much of the passage is above visual range. In autumn, light winds below an altitude of 1,500 m above sea level over the north Sicilian Channel stimulate departure from Sicily. If passage is in progress ahead of an approaching cold front, it ceases, and the birds make for land. As already mentioned, fronts in the Mediterranean are frequently accompanied by heavy rain and thunder. Though it has been claimed that soaring birds use the upcurrents in thunderclouds to cross this part of the Mediterranean, it is extremely unlikely, since the birds probably attempt to avoid the potential danger of the accompanying strong downdraughts. There is also little passage in the unstable air to the rear of a cold front.[170] Raptor migration over Malta, independent as it is of thermals, may be quite frequent at night. Large influxes are recorded early on mornings when strong winds blow from the southeast in autumn, or northeast in spring, particularly when overcast. This suggests a departure in anticyclonic weather, followed by a certain amount of drift and landfall ahead of an approaching depression. Like many raptors, Honey Buzzards are highly gregarious on migration, and it has been suggested that the concentration of migratory flocks assists orientation.[171]

Vultures, eagles and buzzards of the genus *Buteo* typically avoid wide sea crossings. In North America, migratory raptors follow north-south mountain chains in vast numbers. There is an anecdote concerning a party of rather unfortunate Turkey Vultures which must have glided unknowingly out over fog-covered sea, only to find no upcurrents to sustain their flight. The flock of 55 became exhausted, and descended through the fog, where they alighted on a boat in calm conditions and a visibility of about 800 m. When the fog eventually cleared to reveal the coastline, the vultures took flight, but were so exhausted that some fell into the sea and drowned.[108]

Migrant Golden Eagles and Buzzards from Europe are rarely observed in Britain, since the North Sea presents a considerable barrier. Populations of the Rough-legged Buzzard fluctuate according to prey availability, and there have been notable October falls of birds in eastern areas. Some arrivals are in unstable polar airmasses with fresh to strong northeast winds, and apart from a tail wind component, it is possible that the birds may gain an element of assistance from oversea thermals.

The crossing of the narrow straits at Gibraltar and the Bosphorus is undertaken typically on warm sunny days with light winds. Migration is curtailed by continuous rain, low cloud and poor visibility. Short-lived showers have little effect, but strong winds influence the track of most birds in some way or another. In strong northeast winds at the Bosphorus,[141] and in strong easterlies at Gibraltar,[50] passage is reduced markedly by drift, with the main migration further downwind. The only species to pass in large

numbers is the Honey Buzzard, suggesting that there is considerable passage over the sea in the upwind direction. Indeed, at Gibraltar when the migration is at its peak in westerly winds, one can see Honey Buzzards flying over the Mediterranean to the east – some mere specks in the distance. The densest raptor passage invariably occurs at the downwind end of the Strait of Gibraltar. Raptors approach from the north in autumn on a broad front, making use of orographic upcurrents over the Spanish mountains, and thermals when available. In prolonged easterlies, some raptors (particularly the strong-flying Booted Eagle) coast into the wind and look for a resting place. They are reluctant to cross the sea and be drifted towards the wide gulf to the west. Some birds do indeed arrive on the Atlantic coast of Morocco from the northwest following such drift. The drift is always most noticeable on the destination shore, since little compensation is achieved over the sea. Thus with east winds in spring, little migration passes over Gibraltar – the main concentration being further west. At the east end of the Strait, the large Short-toed Eagle is more numerous in westerlies in spring than in autumn, being especially prone to drift.[50]

To interpret raptor passage over narrow sea crossings, one needs the full picture over the whole of the area likely to be crossed, and even then, the change of wind direction with height may render conclusions invalid. Migrants continually seek optimum soaring conditions. This in itself makes movements complex and, in certain areas, assessment of wind drift difficult. Winds in the Strait of Gibraltar are determined very largely by topographical features such as the Sierra Nevada in southern Spain, and the Rif mountains in northern Morocco, and are almost always from the east or west. The mountain chains funnel the wind through the Strait, and their influence is marked in both unstable and stable airmasses. At Gibraltar itself, light to

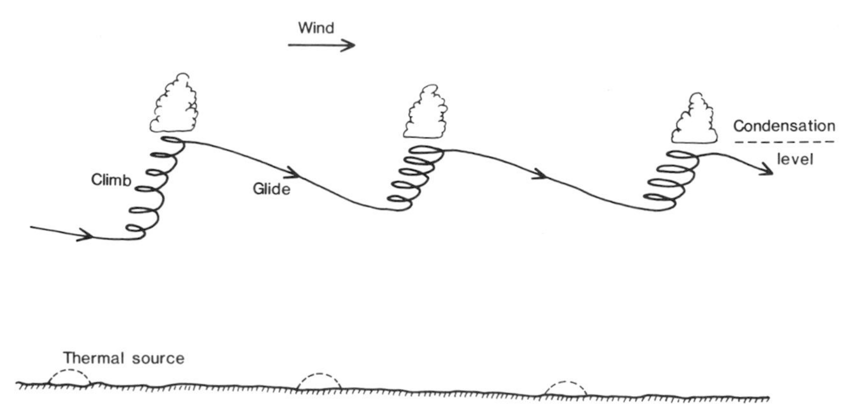

Fig. 33 Cross-country travel by soaring birds with following wind (horizontal scale greatly compressed). (See also Fig. 9.)

moderate easterlies often change to a light southwest or west wind above 1 km, but when strong, this change is at a much higher altitude – occasionally as high as 3 km. Surface wind changes can be very marked and sudden. The altitude at which soaring birds fly is dependent upon the strength of the orographic and thermal uplift over land, and when flying over the sea, they may descend to very low levels. Since winds frequently decrease in strength with height, more drift is experienced at low level. Very tired northbound birds approach Gibraltar just above the waves, but once the cliffs are reached, flapping ceases as they rise on the updraughts and finally the standing wave which is a feature of the 426 m – high Rock itself. In optimum weather, large numbers of birds are so high as to be hardly visible, and easily achieve the opposite shore without significantly losing altitude. High altitude flight is more frequent in the afternoon when thermals are at their most intense, although thermals may be limited by a subsidence inversion within easterly airflows (Fig. 26).

Long distance flight using thermals was referred to in Chapter 2. Large soaring birds may achieve a glide ratio (horizontal distance travelled against height of climb) of 12 to 1 – illustrating the success of this method of travel (see Fig. 33). Migrants locate thermals by observing other soaring birds, and hundreds of individuals may use one thermal in an enormous spiral. In moist thermals, the cloud base may be reached before gliding to the next thermal. Most birds break off from the climb at this point. Since the base of a convective cloud (the condensation level) rises during the day in conjunction with the climb in surface temperature, the maximum altitude reached by soaring birds rises in parallel. There are nevertheless several records of raptors penetrating the base of convective cloud, either gliding out shortly afterwards, or disappearing from view. Two suggestions have been offered for this behaviour. One is that altitudes above the cloud base need to be achieved in order to cross mountain ranges, and the other is that the birds may need to reach the cloud top to orient visually. As long as the vertical currents are not too intense, however, and remembering the constraints imposed by super-cooled water (Chapter 7), a soaring bird would benefit merely by gaining extra height so that the gliding phase can be prolonged.

SUMMARY

This short chapter describes the migration of soaring birds, whose unique weather-related movements are linked closely to their soaring behaviour. Typical species, storks and broadwinged raptors, follow traditional routes to make use of reliable sources of vertical air currents, and their migrations are almost totally diurnal. Long sea crossings are avoided, and this behaviour is compared to that of raptors which show less dependence upon upcurrents.

CHAPTER ELEVEN

Extreme weather

Mortality during long distance migration is no doubt high, although no figure can be placed upon it. It is probable that 50% of migrants are lost between the autumn departure and the spring return to the breeding grounds.[107] Mortality of resident and short-distance migrant populations may be equally high, and it is perhaps opportune now to examine the wider aspects of extreme weather conditions and how they affect avian populations. Considerable changes in these (both in numbers and distribution) are brought about by prolonged periods of winter snow and ice, as well as shorter but exceptional spells of other meteorological elements.

SYNOPTIC SITUATIONS

Within the continents of Eurasia and North America away from the ameliorating effects of the mid latitude westerlies flowing from a relatively warm ocean, periods of prolonged winter snow and ice are normal, except that arid regions do not experience deep snowfall. Bird species which remain during winter are eminently adapted for survival in such climates. The milder and windier western seaboards support a larger variety of species, but temporary changes from an essentially oceanic-type to a continental-type climate can have far-reaching effects for these. In northwest Europe, one such modification is due to a reversal of the prevailing southwesterly airflows over the eastern North Atlantic. A situation giving a drier flow from directions between northeast and southeast (i.e. from a cold, often snow-covered landmass) will cause temperatures to drop considerably. In Britain, the North Sea takes the extreme chill from the air, but serves also to add moisture. Such an airflow over Britain occurs if high pressure lies over Scandinavia and the Norwegian Sea, and depressions are confined to lower latitudes to the south or southwest. Since winter anticyclones are persistent over the landmass of northern Eurasia, with ridges extending westwards to Iceland and Britain from time to time, easterlies can be prolonged, with low temperatures and variable amounts of snow (usually in the form of showers). Away from the normally cloudier east coasts, this results in frequent frosts but relatively cloudless conditions. With existing snow cover, nocturnal temperatures fall very low.

Two particular situations give rise to widespread and deep snowfall. The first requires an incursion of warm moist frontal air from the south or southwest, with colder air beneath it to the north and east preventing the melting of falling snow. If a ridge of high pressure acts as a block over or to the northeast of Britain the precipitation will persist as snow, with the low level warming of the air remaining to the south of the slow-moving frontal systems. The extent of the northward movement of warm air determines both northern and southern limits of the snow, and ascent over high ground accentuates the intensity of the snow fall (Fig. 34).

The second notable snow situation is that in which unstable northerlies or northeasterlies cover Britain to the rear of a depression which has travelled east to stagnate over the continent. The cold airflow, moving as it does over relatively warm seas, is unstable and produces heavy snow showers. The clear periods between showers may be interrupted by minor disturbances in the airflow which give longer periods of snow. These may take the form of nonfrontal depressions known as 'polar lows'. The extent of snow showers over the land is determined by the precise direction of movement, with high ground blocking their penetration. Since these situations normally result from depressions whose penetration into Europe is substantial, they are not as persistent as the 'blocking' pattern in the first example. However, if extensive snow is followed by an anticylonic interlude or merely clear skies in the cold air, frost conditions almost as severe may arise (Fig. 35). There are other combinations of pressure and frontal systems which give more localised cold weather, but the persistence is invariably dependent on the degree of blocking. More mobile situations usually provide milder intervals.

SEVERE WINTERS

Exceptionally severe winters in Britain during the present century have

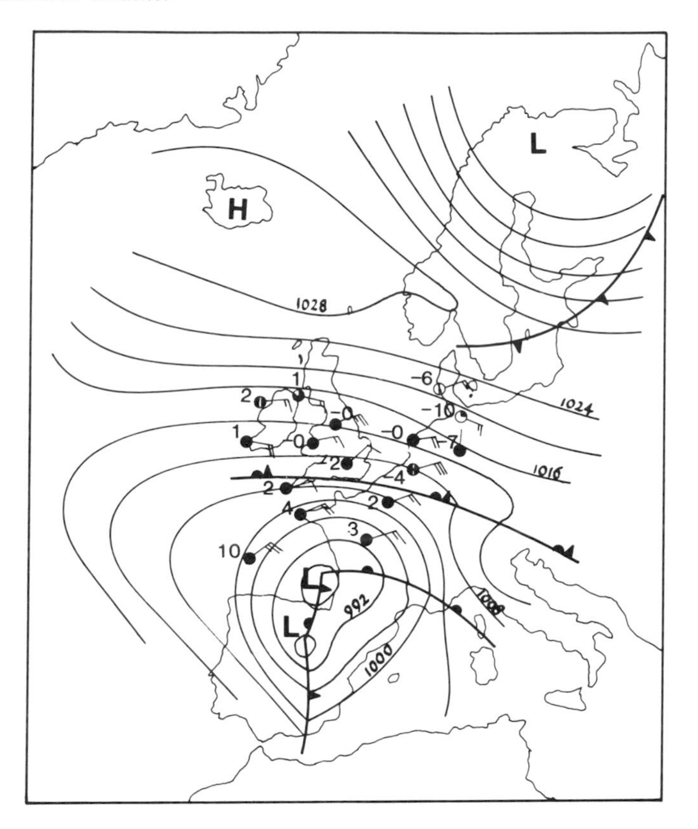

Fig. 34 Easterly weather with snow, 1200 hours, 2nd January, 1963. Widespread snow fell over central and southern England the following night. Very low temperatures (°C) existed in easterlies over the near Continent.

been those of 1916/17, 1946/47, 1962/3 and 1978/79. There is no evidence that the 16 year cycle of the last three is due to anything other than chance.

The winter of 1962/63 was the coldest in central England since 1740. The weather pattern was markedly anomalous during the ten weeks from late December 1962 to early March 1963 (Figs 36, 37). A high pressure anomaly extended east and south to cover most of northwest Europe, while over the Barents Sea, depressions drove intensely cold airmasses from the Arctic south into Europe and then west into Britain. On the southern flank of these persistent easterlies, depressions approaching from the Atlantic were blocked by the high pressure, but nevertheless were able to push frontal systems northeast to give periods of prolonged snow (Fig. 34). Duration of continuous snow cover varied from 10 days on coasts in southwest England to a more general 40 to 70 days from central England to southern Scotland, but even so, the winter was less snowy and more sunny than in 1946/47.[153]

The winter of 1978/79 ranked the coldest since 1962/63, and has been compared to that winter, but it differed in one significant respect; the continuity of frost and snow was broken by several short, but important, periods of mild weather, leading to an absence of reports of dead and dying birds. The very cold spells were accompanied by frequent north and east winds, and considerable snow fell in eastern areas.

In 1963 many rivers which normally remain unfrozen froze so severely that large masses of ice were brought down to the sea during the thaw. Sea

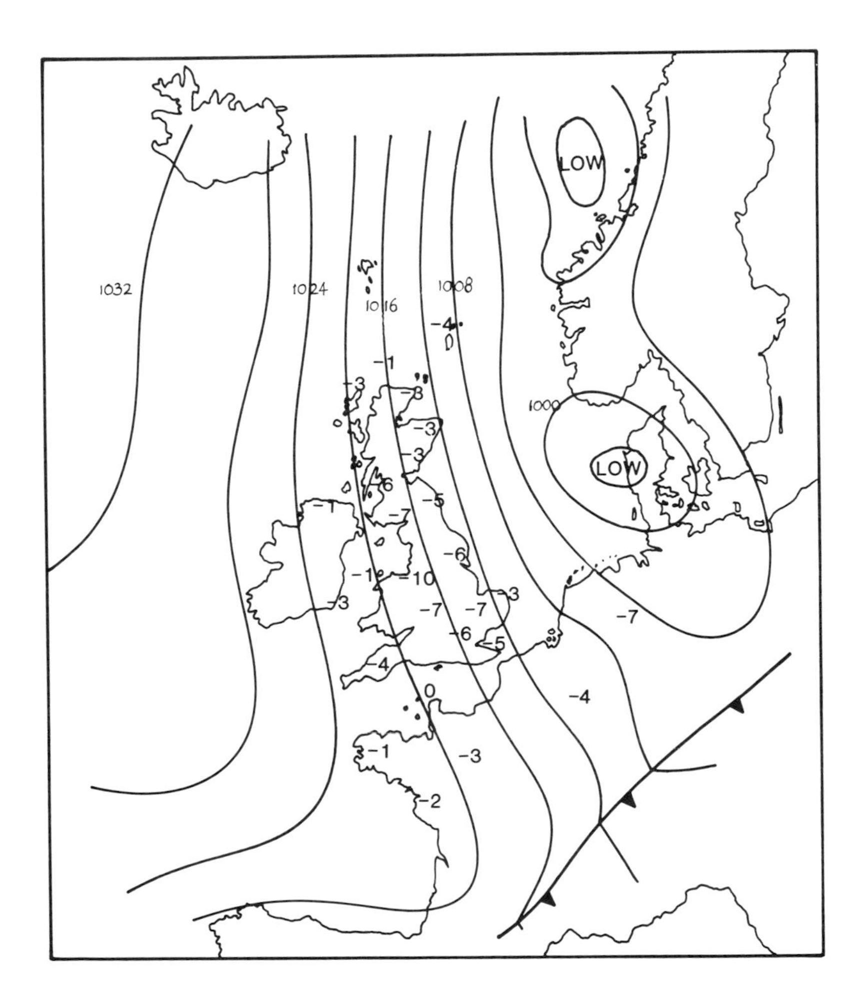

Fig. 35 Northerly weather with snow, 8th February, 1969, 0600 hours. An arctic airmass covered the country. Heavy snowfall had occurred on the cold front as it passed southwards. Further snow fell in showers in places exposed to the northerlies, and cloudless skies in western districts of England allowed temperatures (°C) to fall very low.

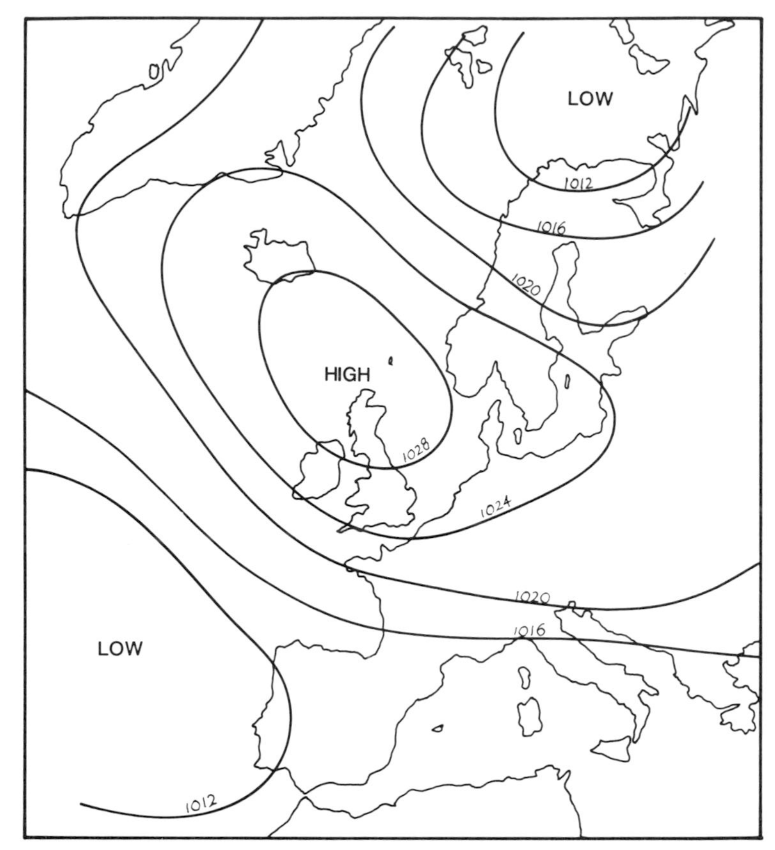

Fig. 36 Mean sea level pressure, January 1963. Compare with normal, in Fig. 47.

temperatures in the eastern English Channel fell below 4°C (some 4° to 5°C lower than average) and the prolonged easterly winds prevented the flow of warmer water through the Straits of Dover (Fig. 38). The sea froze in the Wash and Thames estuary, with frozen shores and land-fast ice from Hampshire to northeast England. Shore and sea ice were also severe in northwest England and in the Severn estuary where intensely cold air flowed westwards from the land. In the River Severn ice floes were several metres thick in places, and large areas of floe ice – one about 10 km × 4 km – lay off Kent. The low North Sea temperatures had been compounded by the previous poor summer in Scandinavia and by the cold winter of 1961/62.

COLD WEATHER MOVEMENTS

When food shortages result from extensive snowfall and severe frost, mobility enables birds to move away from the extreme conditions. Many species, especially the seasonally gregarious, make movements to avoid starvation – often over a long distance. Wynne-Edwards called these movements density-adjustments,[195] in which a proportion of hardier individuals remain at a lower density with less competition – with a reflux on the return of mild weather, though some may not return until the season ends.

There are particular species that exhibit some form of hard weather movement in almost every spell of cold weather. These include some plovers and thrushes and Skylarks, all normally open country feeders. Movements in the 1962/63 winter took place on a massive scale in north and west Europe.

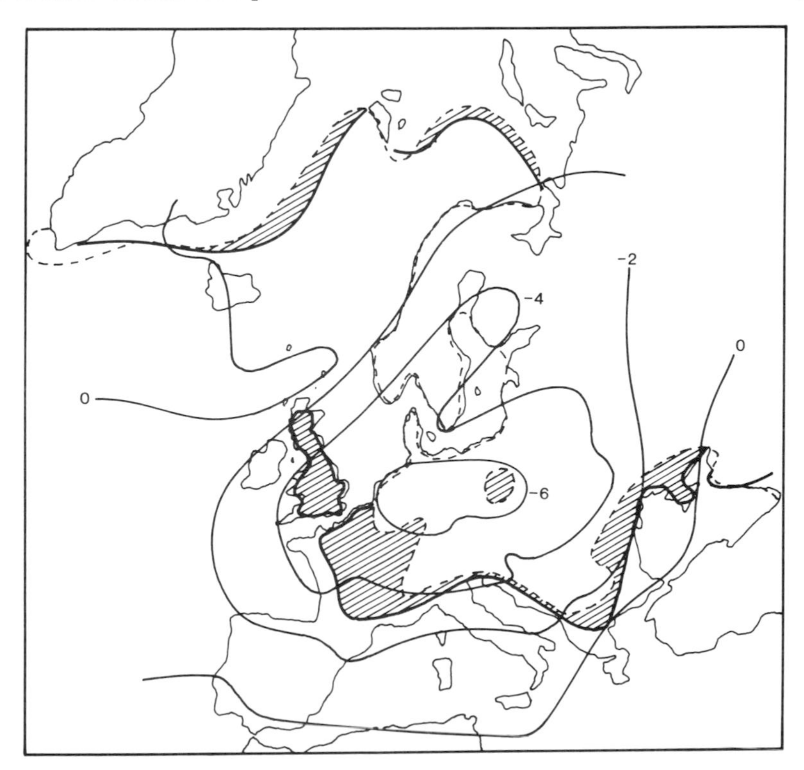

Fig. 37 The winter of 1962/63 – snow and ice distribution.
________ Surface air temperature anomalies (°C), December 1962 to February 1963.
------ Normal position Arctic ice limit, end January. Normal positions of 1 inch (2·5 cm) snow depth limit, 31st January.
▬▬▬ Position of Arctic ice limit, end January 1963. Position of 1 inch (2·5 cm) snow depth limit, 27th January, 1963. Baltic sea ice was at its maximum in late February.
////////// Excess snow/ice (from Murray 1966)[109].

Many British populations moved out and there were abnormal influxes into built-up areas and on to unfrozen shores. Much of the mortality within Britain was probably of immigrants from Europe. That the mortality was not as large as in previous very severe winters was partly attributable to the suitable conditions for emigration, with periods of clear skies in the easterlies, and relatively mild weather in the extreme southwest of Britain at the onset of the snow.

Observations by radar during that winter showed countrywide movements. Many flocks moving from frozen areas ran into appalling snow storms which forced them to fly low.[34] Poor visibility inhibits hard weather passage and the avoidance of snow showers produces erratic movements. Wind direction may be used as a 'signal' to move away downwind from approaching poor weather.[89] Since the direction of surface winds often differs from that of the movement of the disturbance, this may prove misleading and serve to bring the birds into poorer conditions. Radar has shown that midwinter movements in a westerly or southerly direction over the southern North Sea

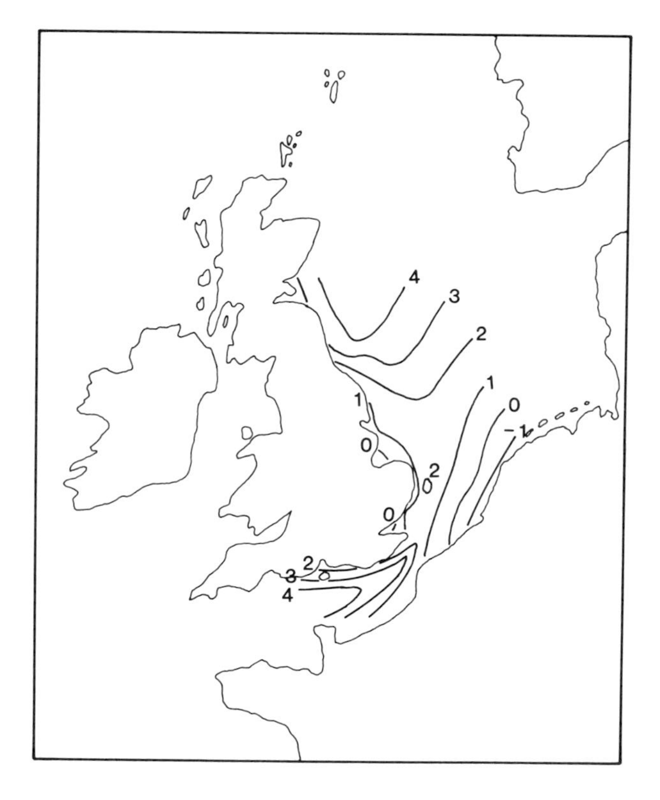

Fig. 38 Five-day mean sea surface temperatures (°C), 24th to 28th February, 1963 (from Tunnell 1964)[175].

Skylarks flying from hard weather

may not only be initiated by low temperatures, but also by winds with a marked east or north component respectively. Return movements are stimulated by winds from the west and south, and thus birds generally move downwind.[85]

Wildfowl wander from their normal wintering grounds in cold weather. Whooper Swans wintering in Iceland fly south at any time if conditions become too severe. Bean Geese wintering in southern Sweden remain inactive but depart if a thaw does not take place within about three days. If the cold weather is accompanied by blizzards, departures take place at the first lull or even before the storm. Since geese typically use traditional wintering grounds, hard weather movements may force them into unfamiliar regions in which difficulty is found in locating suitable feeding and roosting sites. Other examples of wildfowl movements in Europe include those of Smew into the North Sea basin and Britain, of Teal to France and Iberia, and of Pochard to southern France. The last two species (and one or two other ducks) show interesting sexual differences in which the large drakes can tolerate colder weather than the ducks, resulting in earlier cold weather passage of the latter and a more northerly winter distribution of the former.

During the winter of 1978/79, large influxes of water birds were recorded in Britain, chiefly Smew, Goosander, Red-breasted Merganser and Red-necked Grebe, but also other grebes and divers.[23] The weather at the time of the influxes in January and February was snowy with strong cold easterly winds; the synoptic situations were similar in pattern to that illustrated in Fig. 34, except that easterlies extended much further into the central Baltic. The differences in species composition of the separate influxes suggested that the initial distribution of the various species plays an important part in the timing of their movement. This distribution itself is partly due to past weather in the source areas, and it is probable that a contributory factor was the extensive sea ice around the normally ice-free Danish islands.

The most notable of waders to make weather movements are inland birds, particularly the Lapwing and Golden Plover. Both species were recorded in large numbers in Morocco at the onset of the snow of late 1962. Their arrival coincided with one of that country's wettest periods on record (lying as it did in the unusually southern depression track) and consequently ideal feeding conditions prevailed. Marked movements of Lapwings also occurred in the cold spell of early 1979, when more British-ringed birds than normal were recovered in France and Spain. Movements of Lapwings into southwest Europe are frequent enough for the Spaniards to call them 'Avefria' – the bird of the cold. I watched flocks totalling thousands passing southeast over Gibraltar in early December 1973, after a persistent ridge of high pressure from central France to central Europe had given cold northeasterly winds over southern France and northern Spain. In some winters birds reach the Atlantic islands of the Canaries and Madeira.

One of the most dramatic hard weather flights on record is illustrated by the recovery of a Warwickshire-ringed Redwing on a ship 1000 km NNW of the Azores within three days of ringing in early January 1963. Low level winds indicate that it could easily have travelled this distance in 36 hours, revealing not only the extent and persistence of the easterly winds, but also giving an indication of the mortality which must have ensued in species fleeing westwards from potential starvation. In exceptional circumstances, hard weather movements carry birds on a complete transatlantic crossing. Lapwings are particularly prone to being swept from Europe to North America, and two notable invasions took place in December 1927 and January 1966. The birds were undertaking a cold weather movement in the British Isles, moving westwards at a time when strong east to northeast winds existed right across the Atlantic to the north of an extensive area of low pressure. One bird recovered in Newfoundland in 1927 had been ringed as a nestling in Cumbria. Another record – usually quoted in the context of range expansion – is that of Fieldfares reaching Greenland. In January 1937 wintering Fieldfares were driven out of Norway by cold weather, only to be carried by southeasterlies into polar regions. Birds were found dead on Jan Mayen, Iceland and Greenland; ten years later a flourishing colony was discovered in southern Greenland (see also Chapter 8).

The propensity of some species to make substantial movements contrasts with the immobility of others. For example, British Wrens perform only short distance random movements, and an intimate knowledge of winter territories may have an important survival value. Other British residents move coastward, and short distance movements also often show a southwestward orientation towards the milder Atlantic coast and offshore islands. The oceanic climate of Ireland affords a regular haven, and enormous numbers of hard weather migrants appear immediately cold weather sets in further east, with an emigration as soon as mild weather returns. In contrast, mild winters result in little movement, and may even give rise to an abnormal distribution of wintering birds by delaying (or stopping altogether) the exodus of water birds from normally ice-bound seas and lakes.

POPULATION CHANGES

It is clear from the foregoing discussion that avian populations can be reduced by severe weather in areas where they are not fully adapted to tolerate it. However, with the tendency of some species to move to a more optimal environment, it is not surprising that certain populations show no major fluctuations in size unless (of course) the emigrants cannot find suitable alternative wintering areas. Even the species that do not emigrate are able to recover within three to five years. In this they are assisted initially by a decrease in competition for food and space. Should there be more frequent severe winters then these populations will take longer to recover, and the species most at risk are those at the edges of their range where their existence and density are largely dictated by the environment in the form of temperature and shelter.[195]

Turning our attention to British breeding populations, we are fortunate in that important annual censuses are undertaken, notably the Common Birds Census (CBC) and Waterways Bird Survey of the British Trust for Ornithology. These surveys, together with monthly counts of wintering wildfowl and waders and other research tools such as ringing, have shown quantitatively the effects of extreme weather on both resident and migratory species.[197] Since the inception of the CBC in 1962, the most severe winter was that of 1962/63, although its effects have since been repeated to a lesser degree. During that winter, most of the 15,000 corpses found were understandably of larger birds, with Woodpigeon, Starling and Lapwing being the most prominent. Small species such as tits, Treecreepers and Goldcrests died unnoticed, and were only conspicuous by their absence. However, they suffered less than in 1946/47 since there were fewer spells of freezing fog and glazed ice. A survey of the 1962/63 winter[34] found that, of the 23 worst affected species, ten were dependent upon water and moist ground for feeding, two were very small (Goldcrest, Wren), two were at the edge of their range (Bearded Tit, Dartford Warbler), and eight of the others were species commonly affected by hard weather. The CBC showed a 78% reduction in breeding strength for the Wren and 75% for the Mistle Thrush, while Song Thrush, Green Woodpecker, Lapwing, Pied Wagtail and Moorhen had all decreased by over 50%. The figures compared favourably with less precise estimates for previous severe winters, although comparisons are not necessarily valid since there may be marked differences in the availability of food (particularly seeds and fruits). Smaller reductions were found for the Skylark, Blackbird, Robin, Linnet and Reed Bunting, while some finches and buntings increased – reflecting differences in type of food (Fig. 39). Some species decreased

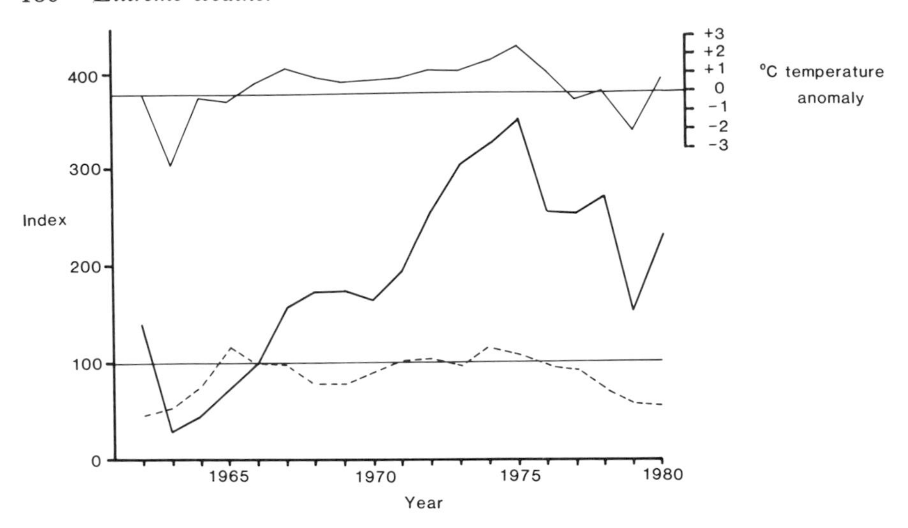

Fig. 39 Index of breeding populations on farmland (1966 = 100), 1962 to 1980. ———— Wren; – – – – Bullfinch. Also shown is temperature anomaly for previous winter for England and Wales (December to February based on 1941–70 mean).

surprisingly little when a reduction would have been expected, but since the winter of 1961/62 had also been poor, populations may already have been at a low ebb. In this case, competition would have been lower in the second winter and populations more viable. A riparian bird not monitored in the CBC, the Grey Wagtail, also suffered serious losses during the winter in question; an analysis of data from various sources showed a fall of 80% from a high population in 1961 to that of 1963.

The effect of food abundance is illustrated very clearly by changes in the Bullfinch population in the Oxford area in 1963 and 1964. Due to the massive stocks of Ash and other seeds in the 1962/63 winter, and their continued accessibility, mortality was only 33%. Since fruiting does not occur at a high rate in two successive years, stocks in 1963/64 were very poor and, despite a mild winter, mortality rose to 68%.[87]

By 1966, many species had recovered to their pre-1963 levels or even shown an increase, but those affected most were still at a relatively low level. Of course, there are other factors which may inhibit recovery, and as in all censuses there are inherent difficulties in assessing population changes. When a population is low it may be that only the most favourable habitats are occupied, with the poorer ones untenanted. These latter are re-stocked gradually from the favourable areas as the population recovers. The Wren is one such example; its reduced population after 1963 was concentrated in wooded areas and in waterside vegetation, but as it increased, so gardens, orchards and hedges were re-colonised.[190] The Goldcrest is a typically

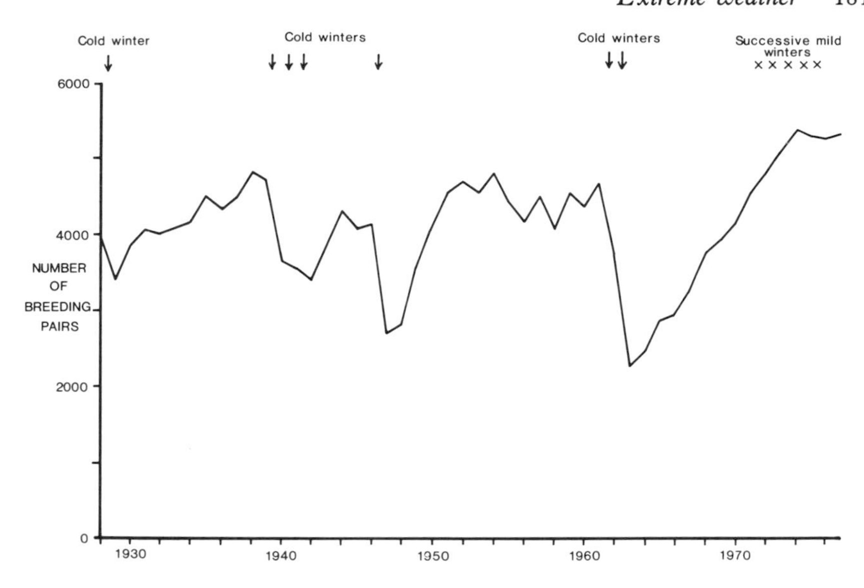

Fig. 40 Grey Heron population levels in England and Wales, 1928 to 1977 (modified from Reynolds 1979)[144].

coniferous woodland breeder, but in lowland Britain will move into deciduous woodland when the population is high after mild winters. After the winter of 1970/71, Goldcrests showed a massive 48% increase in woodland censuses, spreading into areas where they were not normally found. Yellowhammers colonise woodland when numbers are high, and mild weather in November is particularly important for survival, resulting in high spring populations; after a cold start to the winter, there are fewer breeding birds in the following spring. The reduction of population pressure on farmland by the colonisation of woodland raises breeding success.

The resident Dartford Warbler is very susceptible to severe cold weather, and crashes in the population occur regularly. During the two consecutive cold winters of 1961/62 and 1962/63, about 98% of the British population was killed. Such residues offer only limited scope for recolonisation, though healthy recoveries will result from a succession of mild winters. The New Forest population rose from a few pairs in 1963 to 250 pairs by 1974 after the mild winters of the early 1970s; and the total British population had risen to over 400 pairs by 1977. However, it fell to 100–200 pairs by 1978 when there was glazed ice in the February. The mild winters of the early 1970s were also reflected by a peak in records of wintering Chiffchaffs in Scotland in 1975/76, after which numbers declined.

An interesting effect of severe winters upon a summer visitor concerns the Pied Flycatcher, which in Britain competes for nest holes with tits and Nuthatches. Its late arrival puts it at a disadvantage in the search for nest sites, though for several decades now the provision of nest boxes has reduced such competition. After the winters of 1916/17 and 1946/47, an increase in population was noted while its resident competitors were at a low ebb.

Resident species dependent upon water and marshland show fluctuations in sympathy with winter conditions. The Grey Heron, a bird censused annually in England and Wales since the late 1920s, is typical of these (Fig. 40). After 1946/47 the population had been reduced by 35%. A recovery followed, but in the early 1960s the numbers fell again, this time by 52% over two years to a minimum of 2,250 nests. By 1973 there were 4,925 recorded nests, but the recovery was slow. Herons had recovered within two to three years after previous crashes, and the slower increase after 1963 was tentatively attributed to the effect of toxic chemicals. Its relative, the Bittern, suffered heavy mortality in Europe in 1962/63, even greater than after the cold winter of 1955/56 which resulted in the deaths of between 11% and 23% of the Dutch population. A survey of British records of wintering Bitterns revealed that, though mortality rises in severe winters, dead birds are more likely to be immigrants since no significant overall changes occur in the British breeding population.[10]

Reductions in populations in winter affect the breeding stock. Weather poor enough to reduce populations in summer is very rare. An exception is found in high latitudes where arctic wildfowl and waders may regularly suffer from very low breeding success and, although mortality of adults may be low, little recruitment occurs. This can often be seen in the juvenile to adult ratio in wintering flocks, both in Europe and North America. Hugh Boyd[17] has pointed out that Siberian White-fronted Geese are apparently adapted to an environment in which large-scale breeding failures are of regular occurrence. The problem is one of delayed thaws, low temperature and heavy summer precipitation, and for nine of the winters between 1955 and 1976, the proportion of young in flocks of the Siberian form of the Brent Goose was less than 10%.[129] However, the cyclical nature of this subspecies' fluctuations has recently been attributed to predation linked to lemming abundance.

Meteorological elements other than those associated with low temperature are also responsible for changes in size and distribution of populations, operating through food shortages and habitat destruction. Of the more notable are wind and drought.

Wind damage to habitat is rare and normally confined to the destruction of vegetation. Distributional changes in damaged woodland were highlighted by Dierschke,[33] who had been conducting censuses in a study area on Lüneburg Heath in West Germany prior to a very severe storm in 1972. During this storm violent NW winds to the rear of an intense depression reached gusts of 42 m/s, destroying 10–20 million cubic metres of forest.[100] Most of the damage in Dierschke's study area had been in pine forest, of which 44% was almost totally felled. Only a few single trees or groups of trees remained standing, with a ground layer of heather and cranberry covered by tree trunks and uprooted stumps. Fourfold increases in ground and shrub

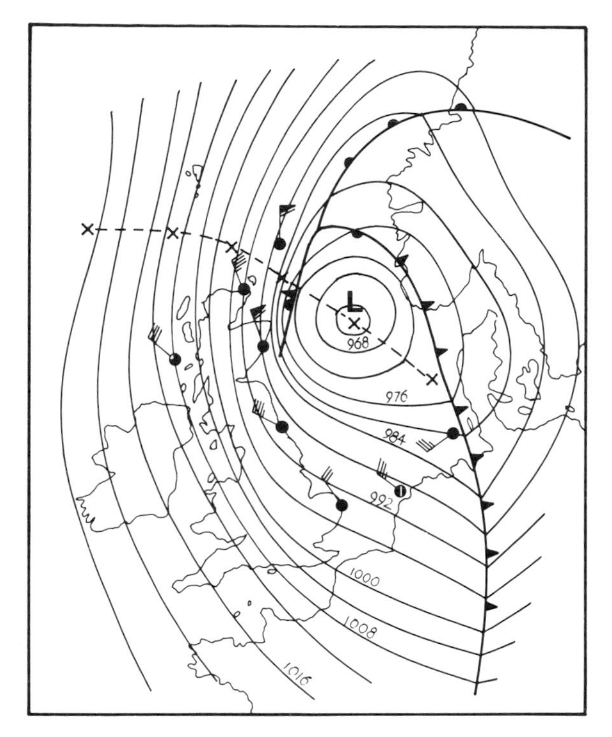

Fig. 41 Habitat destruction in Britain, 31st January, 1953, 1200 hours. Track of depression is shown with positions at 6-hourly intervals. Note extremely tight pressure gradient in northeast Scotland with 30 m/s (60 knots) winds.

nesting species occurred in areas where pines were extensively felled, particularly of Wren, Dunnock, Willow Warbler and Yellowhammer, while arboreal breeders were halved in number. Extra food in dead trees raised the Great Spotted Woodpecker population, while increases in nest sites and song posts benefitted Spotted Flycatchers and Redstarts respectively. The most surprising result of the habitat modification was the colonisation of the area by such large numbers in such a short time, helped no doubt by a substantial reserve of birds outwith the study area.

One of the worst storms in Britain this century was that of 31st January 1953, which laid waste to extensive tracts of forest in Scotland and much reedbed habitat in eastern England and the Netherlands. Hurricane-force NW winds associated with an intense depression over the North Sea coincided with high spring tides (Fig. 41). The combination of high winds and low atmospheric pressure created a surge of water which breached sea defences and flooded over 800 km^2 of English countryside. Although the exact toll of bird habitat will never be known, 6% of Scottish heronries were damaged in the winds, and in East Anglia, the Bearded Tit population fell by 44%

Dartford Warbler near gorse fire

between 1952 and 1953. Colonisation of new sites led to a rapid recovery of the latter species in 1954.

In recent years drought, the prolonged deficit of rainfall, has been a problem for man in many areas of the world, but also for birds through desiccation and fire damage to habitat. Natural fire destruction depends upon previous rainfall, temperature, humidity and wind velocity and in western North America lightning-generated fires account for 80% of the area burned. Such fire-affected ecosystems often hold a higher density and diversity of bird-life. Plant communities consist of species able to recolonise burnt areas quickly from without. One or two avian species, such as the Dartford Warbler, are particularly prone to population reductions due to heathland fires, more so because the fires occur mainly during the latter part of the breeding season when the young are newly fledged. Such fires begin (either naturally or by accident of man) after lengthy dry spells, and were severe in the long drought of 1975 and 1976. Since the latter summer was unusually hot and dry, with only 35% of normal rain falling in England and Wales in three months,[110] desiccation of vegetation and outbreaks of serious heath and forest fires were widespread in western Europe, where the greatest rainfall deficits were found in a belt from Brittany to East Anglia. As with most periods of climatic variability, causes were obscure although a feedback of sea temperature anomalies via the atmospheric circulation may have played an important role (see Chapter 1).

Despite the drought, long term effects on bird populations were small. There was an apparent paucity of birds visiting gardens in the following autumn, notably Wrens, Great and Blue Tits, Blackbirds and Song Thrushes. This was attributed to the lack of food for nestlings and fledglings, but breeding censuses revealed few significant changes which could be ascribed to the drought. The Dartford Warbler was reduced in some areas but in 1977 the species had a good season after colonising unburnt heathland, although the slow regeneration of vegetation may impede recolonisation of former habitat. The long hot summer of 1976 evidently helped some species, and there was a resurgence of breeding of Montagu's and Marsh Harriers, Hobbies and Golden Orioles in southern Britain, at the northwestern and most maritime edge of their breeding ranges.

A significant setback in recent years to the breeding populations of some of Europe's migratory passerines (and even perhaps wetland species) has been a deterioration of their African winter habitat. Large numbers of migrants spend the months of the northern winter in the Sahel savanna (a zone of grassland and scrub lying along the southern border of the Sahara desert) or use the region as a staging post on their way further south. Here aridity alternates with seasonal rainfall. It is a dry area with scanty rain falling in late summer as the ITCZ reaches its northern limit. The lands include the basins of the Sudan sudd, Lake Chad, the Niger Inundation Zone and the Sénégal river – all of which are flooded as a result of greater rainfall over higher ground to the south. There was a period of near or above normal rainfall between 1949 and 1967, but then a drought began, with progressively lower rainfall from 1968 until 1973, and only minor fluctuations since. The drought has been attributed to changes in the vertical atmospheric circulation brought about by anomalies in tropical ocean temperature.[203] Alterations in land surface properties, such as soil moisture and reflectivity of solar radiation (e.g. by vegetation changes) have helped to maintain and enhance the drought, but there is also ample evidence that the effect was accentuated by the negative management of the human and domestic animal populations, which had increased in the higher rainfall period of the 1950s and early 1960s. Man has carried out extensive clearances by burning, often after the rainy season – at the hottest, driest and windiest time when the northeasterly Harmattan wind prevails. Thus a combination of drought and man's mismanagement

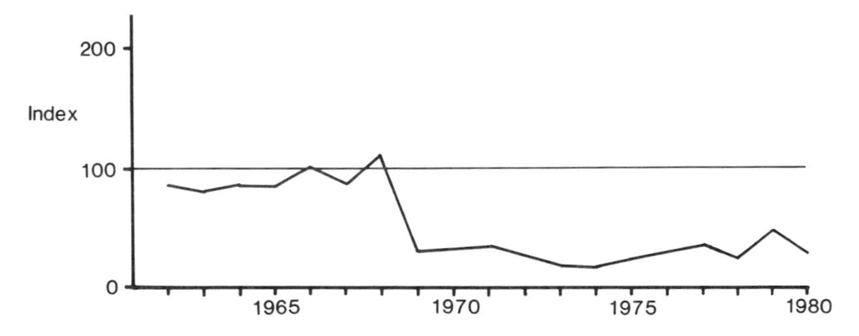

Fig. 42 Index of Whitethroat breeding population on farmland (1966 = 100), 1962 to 1980. Marked drop in 1969 coincided with drought in winter quarters, with no subsequent recovery.

upset the ecological balance of the region, and affected the Palearctic migrants that use the region. In northern Nigeria, the southward encroachment of the desert has also resulted in the invasion of birds from the north, while in other areas local birds do not breed when the food supply fails. The Palearctic migrants arrive after the end of the rains when there are optimum food supplies and, although the habitat subsequently deteriorates progressively through the winter, some species remain in the absence of competition from resident species, while others move further south. High mortality during the drought is thought to have reduced populations through a lack of food, water and shelter. Figure 42 shows the Common Birds Census indices for the most hard-hit species – the Whitethroat – and populations of Redstarts, Sedge and Garden Warblers, Yellow Wagtails, Spotted Flycatchers and Sand Martins have also suffered.[192] Partial recoveries have been effected by recent fluctuations in rainfall but, for the Whitethroat and Sand Martin, the drought remains a relentless problem.

OTHER MORTALITY

I have covered various aspects of mortality, through extremes of weather and food shortages, yet there are other events in which the weather is more directly responsible for deaths. These act to a relatively small degree on population levels (certainly very much less than food scarcity) except on rare occasions when populations may be almost eliminated by unusual weather events, e.g. the effect of tropical storms on resident island species.

A. R. Jennings[75] analysed the causes of death of 1,000 free-living wild birds, and found that just over 10% died as a result of what he termed climatic factors, both direct and indirect. Naturally it is the extremes of weather that directly kill birds. Violent precipitation in any form is dangerous, especially of the intensities found in convective storms. Prolonged cold rain is worse than dry cold, since it may wet plumage sufficiently to reduce insulation, and the battering force of heavy precipitation is often powerful.

Hailstones from the convective storms of some continental or tropical areas can be large enough to kill cattle, and therefore considerable numbers of birds are at risk in such storms. The summer hail storms of central Canada are almost unequalled for their violence and damage. Most frequent in July, two of these storms in 1953 killed an estimated 148,000 wildfowl in Alberta. In the first, over 1,800 square kilometres of parkland were flattened by hailstones as large as golf balls; even emergent vegetation in lakes and ponds was reduced to pulp. After another storm four days later, the land appeared seared with no vegetation, and these storms killed songbirds, hawks, grouse, coot, grebes and ducks. The loss of waterfowl was calculated as 93.4% of the population.[159]

Hail is by no means confined to summer. In November 1973, ducks on a local feeding movement in Arkansas flew into a thunderstorm where they encountered hailstones up to 5 cm in diameter; many were struck and grounded, suffering broken bones, gashes, cuts and bruises. Eighteen birds were found frozen and ice-encrusted; one female Mallard had pieces of ice 2 cm in diameter frozen to the tips of feathers on the neck, breast and flanks. From the descriptions of the icing, it is clear that many birds must have been flung high into the subzero temperature region of a Cumulonimbus, where they underwent the same process as a water droplet during its formation into a hailstone (see Chapter 1). Four dead birds were found with head feathers singed, possibly by lightning strikes.[149] Death by hail is only very rarely observed in Britain, although many fatalities must go unrecorded. Severe thunderstorms and damaging hail occur most often from June to August, in the London area and the southeastern parts of the Midlands, coincident with the highest mean summer temperatures.

Even in winter, however, British thunderstorms are occasionally severe enough to cause mortality. On 3rd January 1978, a squall line ahead of a well-marked cold front passed over eastern England and killed 140 geese, mainly Pink-footed but also a few Canada, Greylag, Brent and Bean Geese. The corpses were found in a narrow 50 km-long corridor in Norfolk, and post-mortems of four showed ruptured livers and haemorrhaging consistent with damage by blast or decompression, and extensive injuries and bone damage indicative of a fall from some height. The squall line was accompanied by several groups of tornadoes, widespread thunderstorms, heavy rain and hail, and was an extreme example of local marked instability and vigorous vertical convection. It was thought that the geese were flighting at dawn near the Wash when overtaken by the storm, sucked up by intense updraughts (perhaps even a tornadic vortex) and subsequently deposited from the cloud as it passed over Norfolk in mid morning.[174]

Circumstances in which birds are unable to find a safe roost in thick fog have already been described (Chapter 6). Fog as a direct cause of mortality may not occur, but there are well-documented records of asphyxiation in smog (a contraction of the term 'smoke fog'). This is a particularly dense fog in which pollutants such as smoke have played an important part in both formation and thickening; it is mainly a winter phenomenon trapped under an inversion. In Britain, the passing of Clean Air Acts in 1956 and 1968 has almost eliminated the incidence of smog, but in 1959–60 there were several instances of mass deaths due to asphyxiation, mainly of Starlings. Although examination of corpses showed internal inflammation and haemorrhaging, and extensive pulmonary damage, asphyxiation may be merely a proximate factor in the mass deaths, the ultimate cause being dependent on the location of the birds at the time they are overcome (i.e. over water or busy roads).

There have occasionally been large numbers of birds killed while flying in fog and low cloud (or in strong winds) by collision with tall structures, such as power lines, lighted towers and oil rigs. This happens most frequently to nocturnal migrants in dense fog or cloud accompanied by precipitation. The refraction and reflection of light by water droplets increase the sphere of illumination and confuse the migrants. Very strong winds are also hazardous to flocks of birds under certain circumstances. Coastal roosting flocks have been blown into the sea and drowned (see Chapter 6). Although birds will seek shelter or flatten themselves against the ground, the sudden horizontal and vertical dynamic pressure changes shake birds loose from perches, especially small birds.

The effects of adverse weather are more pronounced for birds that are deformed, injured, diseased or heavily parasitised, and these individuals are invariably the first to die in a stress situation. Resistance of healthy birds may be lowered by sudden and large temperature changes. There is evidence of a sex difference in the ability to resist these changes; females appear to be hardier, particularly in polygamous species. Prolonged cold and wet weather is liable to cause lung disorders, especially in young birds with poorly developed thermoregulation, while wet conditions increase food supplies of worms and molluscs which often act as intermediary hosts to large numbers of parasitic organisms. Although all birds carry some parasites, the balance between the avian host's physical condition and the number of parasites is subtle. A food shortage lowers resistance and upsets this balance so that heavy infestation and possibly mortality ensue.

A most weather sensitive source of mortality is botulism. A bacterium *Clostridium botulinum* Type C produces a toxin under certain conditions. The toxin may then enter food and is ingested by wetland birds. For the production of this highly poisonous substance the bacterium must be present in mud, either exposed or under shallow water. The bacterium multiplies when the temperature of its immediate environment exceeds 25°C, and the toxin itself is produced at temperatures in excess of 28°C. Enormous mortality in the USA has followed heavy rain or late snowmelt which caused shallow flooding in spring, succeeded by warm sunshine. These conditions are rare in Britain, but a similar environment is created by hot dry weather which reduces water levels so that mud is exposed and shallow stagnant water is extensive. Rotting vegetation assists development of the bacteria, so that ideal conditions were created on the Norfolk Broads in the drought of 1975–76. L. P. Smith[160] calculated indices based on summer temperature and excesses of evaporation over rainfall. He found a strong correlation between the number of days when air temperatures exceeded 21°C (giving prolonged periods of mud temperature in excess of 28°C), drought conditions and the incidence of avian mortality. From 1969 to 1977, the highest indices and mortality occurred in 1975 and 1976. In 1975, 2,000 to 3,000 wildfowl died in Norfolk, with 2,000 gulls (mainly Herring Gulls) in northeast England and southeast Scotland, while other outbreaks occurred elsewhere. An incident on the Coto Doñana in Spain resulted in the deaths of 50,000 birds, and in 1952 some 4 to 5 million birds died in the USA. Although it is mostly wetland species that die, raptors preying on contaminated birds, or even bathing in the toxic waters, may also be killed.[122]

SUMMARY

In this chapter I have described the weather situations that result in extremes. Although prolonged snow and frost are normal winter features of more continental climates, the milder northern maritime climates suffer periodic incursions of very cold air. Some of the wealth of wintering birds in these regions are less well equipped than others to withstand the results of prolonged cold, and in very severe winters, some populations are markedly reduced. Adjustments do occur, in which part of the population emigrates to a more amenable area, leaving a more viable population of lower density in the original wintering area, though some species move out completely. An improvement in the weather initiates a return of the displaced birds. Populations which are reduced by starvation in a severe winter normally recover within 3–5 years, perhaps longer if two or more consecutive cold winters occur. Those species residing near the edge of their range may risk extermination, or at best a catastrophic decline.

Other weather factors which give rise to numerical and distributional changes to avian populations include strong winds and drought, mainly through their effect on habitat, but direct mortality is also caused by these and other extremes.

CHAPTER TWELVE

Seabirds

The influence of weather on the life style of most seabirds is somewhat different from its influence on land birds. Apart from a few species of gulls and terns, seabirds spend their whole lives over and close to the oceans. Even those that nest in the polar regions and remain in adjacent waters in winter are protected by the ameliorating effects of the sea from the severe freezing that limits the food resources of land birds. Perhaps the one meteorological element that modifies seabird behaviour is the wind. I have shown how polar front depressions deepen while crossing the vast stretches of ocean and, in deepening, generate extensive systems of very strong winds. Any birds living on the oceans of mid and high latitudes must be adapted to exist under conditions of high winds and, if not very low temperatures, considerable wind chill and precipitation. Even when breeding, the immediate proximity of the open sea poses special problems regarding exposure. Some aspects of seabird flight were dealt with in Chapter 2, and in this final chapter, I shall concentrate on the singularities of their feeding, migratory and reproductive behaviour.

FEEDING BEHAVIOUR

There are certain seabirds of mid and high northern latitudes that migrate from the severe winter weather to the opposite hemisphere, or at least to subtropical waters; such birds as shearwaters, storm-petrels, skuas and terns. There are others which remain in the stormier oceans throughout the whole year, spending the winter at sea. These include the Fulmar, part of the Gannet population, the auks and some gulls. Yet others, such as the Cormorant and Shag, stay in coastal waters. That populations survive is an indication of their highly successful adaptation to oceanic weather, although there are occasions when prolonged food shortages cause mortality.

Visual observations of feeding seabirds are notoriously difficult to make, and even more difficult to analyse because coastal movements, particularly in winter, form only a small part of the total. Observations at sea are at present fragmentary. Some seabirds take advantage of local abundances of food, relying on sight to follow others which have already located such concentrations. These local food sources may be found in coastal waters or well offshore, usually associated with oceanographical features such as upwelling and mixing of waters of different origin and nutrient content. The birds move on when food becomes unavailable, and the more aerial species often congregate at fishing vessels. At certain times of the year many populations migrate, and it is especially difficult then to separate feeding movements from passage. The onset of poor weather, particularly gales, and its effect upon sea surfaces and prey may displace seabirds and perhaps force them to seek less rigorous conditions in which to feed. Displacement beyond their normal feeding range gives rise to return movements, often into an abating headwind,[15] in which large numbers of seabirds pass coastal stations and give the impression of mass migrations. As there is no way of knowing the precise origin and destination of these movements, one cannot stipulate their nature on every occasion, particularly as some nesting birds of the more pelagic species may feed over 400 km from their nest site. One can, nevertheless, identify some of the meteorological factors which give rise to such movements.

Movements and technique

As in other avian groups, the effect of the weather on a species' feeding behaviour depends partly on its prey and method of foraging, and partly on its physical attributes, namely size, weight, and flight characteristics. Those birds that are entirely coastal are obviously not adapted to the severe conditions found in mid ocean. The Shag, for example, often cannot tolerate very poor weather even in coastal waters. It feeds chiefly on free-swimming fish, and needs suitably sheltered coasts on which to roost. It finds difficulty in feeding when the seas are too rough and the shoals move away. British populations tend to disperse from breeding sites in winter, but in certain areas show eruptive movements which are more prevalent during periods of strong onshore winds. G. R. Potts[142] discovered that large eruptions take place just as winds moderate after blowing for at least 11 to 12 days from onshore directions. He suggested that any movement undertaken during gales would be suicidal, although it is the persistence rather than the strength of the wind that is significant.

In sheltered western coasts of Scotland – where the islands reduce the force of all winds, even the prevailing westerlies – these movements are uncommon. On exposed east coasts, however, there is little or no protection from wind directions between north and south through east. Long distance feeding movements (of over 20 km) of Shags from cliff roosts in northeast Scotland are more concentrated during winters in which the frequency of onshore winds is much higher than average. After two winters with large movements, when onshore winds were frequent, a winter of offshore winds coincided with small movements. The drop in numbers can be attributed to the ability of the birds to feed satisfactorily close to the roosts. In addition, there was evidence of a smaller wintering population, not augmented by birds from elsewhere – themselves having better feeding conditions.

In contrast, the Gannet is probably one of the most capable and resilient of the diving seabirds. It fishes even when sheets of spray are being blown from the surface, but with winds above force 8 (19 m/s) it does not fly much and feeds little. Adults have a more northerly winter distribution than the more migratory juveniles, but both may find food unavailable for long periods because of prolonged gales. Their fat reserves and insulation, however, allow them to pass these spells without undue stress. They feed out to sea over the continental shelf, where there is little shelter from gales and wintry precipitation.[114] Even when breeding, they fish only occasionally in inshore waters. Before laying, birds leaving the Bass Rock colony off east Scotland during and after gales may be forced to spend longer fishing. Fishing success is dependent upon at least some wind, rather than a flat calm. The wind assists flight, and by disturbing the sea surface, may obscure the view that fish have of a foraging bird, while allowing the bird to see the fish. Bryan Nelson[114] suggested that the eye of the Gannet may even have polarising powers. He quotes a record of Gannets successfully catching pollack off Eddystone in rough seas, but in calm waters they were unsuccessful. Manikowski,[99] watching from a trawler in the North Sea, saw most Gannets ahead of depressions, and in the vicinity of troughs and fronts. Numbers increased in active depressions and decreased in the light winds of cols and anticyclones, but Nelson observed that they were often associated with cols or clearing frontal weather.

A more detailed study of the fishing success of plunge divers (which include the Gannets and their relatives, the terns and one or two shearwaters) revealed a relationship between fishing success and wind speed and sea surface conditions.[39] Terns precede plunge dives with hovering, so that to remain stationary, they require more exertion when the wind is light or calm. Under these circumstances, the fish have a better view of the vigorously flapping tern, and take evasive action. An increase in wind and sea state reduces the effort needed to hover, and also the fishes' ability to see the terns – due to the ruffling of the surface. The result is an improved catching rate.

There is an optimum wind speed and/or sea state that benefits a fishing tern. Above this, the bird has increasing difficulty in maintaining an accurate diving trajectory, while the fish swim deeper in a rough sea. On the Northumberland coast, both fishing success and diving rate of Sandwich and Common Terns were higher with moderate seas than with calm seas, but there was no difference in diving rate with wind speed. This raised the point that sea state did not necessarily reflect the wind speed. In coastal waters, an offshore wind has less effect than an onshore wind, and sea surface conditions may not ease for some time after the wind has dropped. It has been suggested that an increase in body weight may mean greater stability in strong winds, so that the large Sandwich Tern is then a more successful fisher than its smaller relations.[40] When fishing is difficult in cold windy weather, sea terns will take insects and beetles on the wing and bring far fewer fish to their young. Even in West African winter quarters, storms delay or prevent terns from leaving the roost for the feeding grounds.

Sea conditions (initially generated by wind) are clearly an important factor in determining the efficiency of foraging, but the importance varies with the feeding method used. Auks, for example, catch their prey by underwater pursuit. They dive from the surface to depths which often take them well below the surface turbulence of a rough sea, and most recent research has found no significant relationship between the rate at which young Guillemots are fed and the weather, including winds up to at least force 6.[206] Much of the food of British Guillemots consists of sand-eels and sprats, with no change of size evident with differing wind conditions.[11] Auks are fast, but not very efficient fliers. The wings are short, although rapidly flapped, and, if unable to feed, birds may be wind-blown. R. M. Lockley[93] recorded that a light breeze lifts and sustains Puffins in flight. He also observed them flying against a 27 m/s wind, when their ground speed was minimal. In a 40 m/s (hurricane-force) wind, none left the water.

Shearwaters may be displaced from their feeding grounds by strong winds. The passage of vigorous depressions probably makes feeding difficult for large numbers of seabirds, and initiates movements towards better conditions. Equally, drift by strong winds results in compensatory movements.[15] Movements of Manx Shearwaters past headlands are commonplace in early autumn when birds have left their colonies but have not begun to migrate in earnest. They are often accompanied by Sooty, Great and Cory's Shearwaters, the two former species 'wintering' in the northern hemisphere. Most of the shearwater passage along the coasts of eastern Britain is in a northerly direction at all seasons, and off Rattray Head in northeast Scotland is made in two distinct weather situations.[46] Large movements occur in moderate to strong north to northwest winds, usually associated with the clearing weather to the rear of a front or depression (Fig. 43); and in onshore winds between northeast and southeast associated with the poor weather on the northern or eastern flank of a depression, when birds are apparently drifted close to the coast. In these conditions, Manx Shearwaters are often driven by, or move

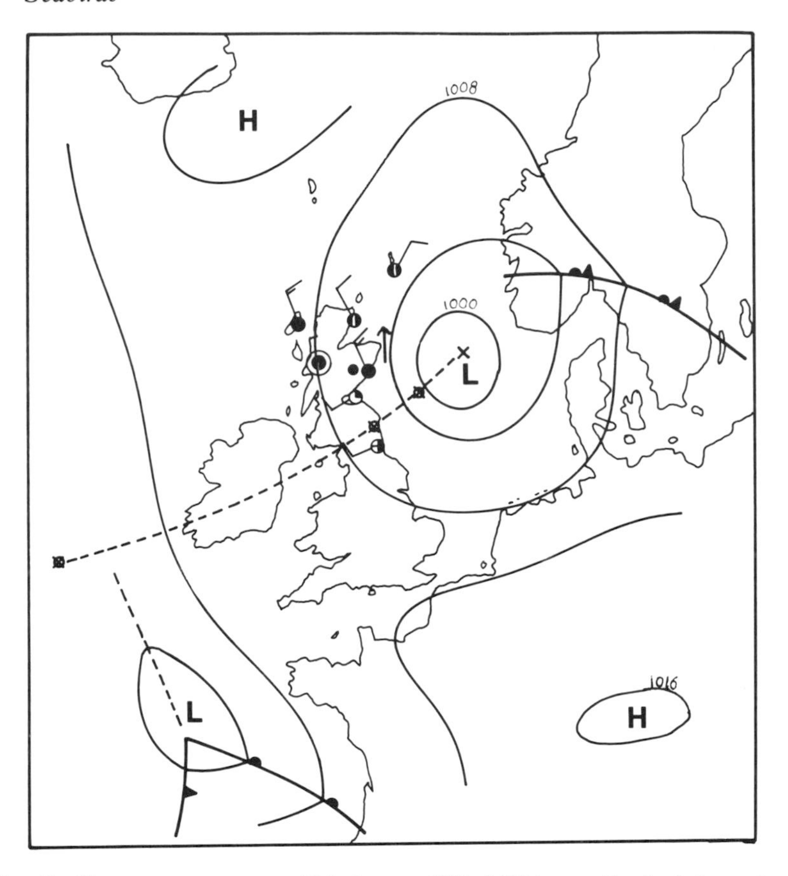

Fig. 43 Shearwater movements 18th August, 1970, 0600 hours. Track of depression shown, with midday positions on the previous three days. The depression decelerated in the North Sea, pulling in strong NW winds to its rear and creating poor feeding conditions for some seabirds. Winds then abated as the depression filled. Northward movements of shearwaters began on 18th August off northeast Scotland (arrowed), with another surge of birds on 21st as a further depression passing to the south brought strong easterlies into the northern North Sea.

ahead of, bad weather into estuaries and firths – they appear right at the head of the Moray Firth in squalls. Movements invariably cease when winds drop.

The first situation is the commoner, and two factors point to the movements being due to feeding rather than migration. First, on 22nd June, 1971, a large northward movement of Manx Shearwaters, accompanied by Fulmars, took place in fresh north to northwest winds behind two consecutive vigorous depressions. These birds were doubtless non-breeders or those associated with the colonies in the northern isles, having been feeding in the North Sea when they were caught in disturbed seas and drifted down wind. June movements such as this were not recorded in 1969 and 1970, and it is perhaps of interest that this particular month (June, 1971) was unusual for its frequency of north to northwest winds in the North Sea compared to the previous two years. Secondly, considerable numbers of Manx Shearwaters migrate during anticyclonic conditions off western Ireland[138] – a weather type that produced no movements at all, at any season, off northeast Scotland. It has been suggested that the steady winds and regular sea wave systems behind many cold fronts give ideal conditions for Manx Shearwaters (and also Fulmars and Gannets) to move to more suitable feeding grounds (Table 5). The picture in the breeding season is naturally somewhat different where coastal sites are close to the route between breeding and feeding grounds. Heavy shearwater passage in summer off southwest England is observed more often from the coast in strong westerly winds, where birds are drifted inshore, and it is probable that feeding birds will follow their prey which itself is affected by strong onshore winds. In calm weather movements take place on a broad front.

Although most shearwaters appear to move out of the North Sea round the north of Scotland, some, chiefly Sooty Shearwaters, pass through the Strait of Dover, possibly after feeding at the southern end of the North Sea, or while on onward passage after entering from the north. Substantial westward movements at Cap Gris Nez are invariably associated with the westerly winds and clearing skies to the rear of a North Sea depression. The abundance of shearwaters off coasts of southwest England and Ireland reflects the exceptionally good feeding conditions over the edge of the continental shelf. The distribution of birds over the shelf at any one time no doubt determines the coastal sites at which they will appear in poor weather and onshore winds, but the true picture cannot be assessed from the land. W. R. P. Bourne[14] saw several thousand Great Shearwaters some 280 km southwest of Lands End in late August 1973. A week later, after a southwesterly gale, a massive passage was recorded at Cape Clear, when 4487 birds passed in five hours. These were birds presumably displaced from their feeding area, and it was just one of the many heavy coastal seabird passages at this site which occur regularly in the poor visibility and onshore winds coincident with the passage of a frontal depression. Prolonged gales may have a cumulative effect by concentrating birds in one area.

Massive Manx and Sooty Shearwater movements in August 1970 in the Irish Sea and off southwest Ireland occurred after rapidly deepening depressions moved east to northeast across the southwest approaches, and the subsequent north to northwest winds initiated a movement into improving weather similar to those observed along the North Sea coasts. Further post-frontal autumn movements have been observed along the north coast of Spain, out of the Bay of Biscay. In these, Manx and Sooty Shearwater passages have shown slight differences in timing in relation to the abatement of gales, with the latter species showing a lower dependence on the post-frontal winds. There is clearly a great complexity in the movement of the various species with perhaps differences in feeding ability in changing winds. Movement upwind to the rear of a depression, and downwind ahead of the centre, quickly takes birds to better feeding conditions, the latter route by circumnavigating the northern flank (movements with a motive analagous to Swift weather flights described in Chapter 4).

Manikowski[99] concluded that Fulmars avoid rough seas. During his observations from trawlers off northeast Canada and in the North Sea, he saw most in moderate winds at the edges of pressure systems, and many fewer in the strong winds of active depressions and the very light winds of anticyclones. Movement to more favourable areas occurred in strengthening winds, but not if strong winds were decreasing or about to decrease – thus suggesting an awareness of future conditions. However, Manikowski did not appear to take the sea state into account in his analyses. Fulmars passing Erris Head in western Ireland stopped moving at the onset of rain, unlike Manx Shearwaters[138] – although the same effect on the latter has been observed at Cape Clear. Despite the extreme midwinter conditions existing in polar seas, Fulmars occur at this season in considerable numbers in the ice-free waters around Jan Mayen and Bear Island. Even in January, they remain when air temperatures are −28°C with fresh to strong northerly winds, though hurricane-force winds killed hundreds one February.

Winter gales from northerly directions often result in massive flocks of gulls in coastal areas of Britain. Herring, Great Black-backed, Glaucous, Iceland and Little Gulls appear on headlands, islands and estuaries, having been prevented from feeding at sea. In Manikowski's studies,[99] Glaucous Gulls were most frequently seen at sea off northeast Canada during quiet anticyclonic conditions. Being coastal species, they remain inshore during stormy weather. Other gulls avoid areas with poor visibility associated with precipitation and fog, and tend to remain ashore. This may be related to the inability to find food or the lack of information from other seabirds regarding food sources.

Immense numbers of Kittiwakes pass northeast Scotland coasts in gale-force winds. The wind direction during these movements, however, is not always onshore, and many may merely be moving to more sheltered feeding areas rather than being drifted on to a lee shore.[46] There is evidence that Kittiwakes fly overland to avoid gales, again to seek more sheltered feeding grounds.[152] Movements of thousands were seen in November 1973 and again in October 1974 passing westwards up the Firth of Forth, and the observer assumed that they could only be crossing to the west coast. These two movements occurred in identical weather situations – very strong north-westerly airstreams behind a depression entering the North Sea, with a ridge of high pressure building west of Scotland (Fig. 44). At sea in winter, Kittiwakes avoid active depressions and frontal systems, and in strengthening winds move to more favourable areas in very large numbers.[99] In the North Sea, these movements are undertaken particularly with westerly winds, and fluctuations in numbers occur very rapidly. They appear to remain in strong winds only if these are abating. Of meteorological factors,

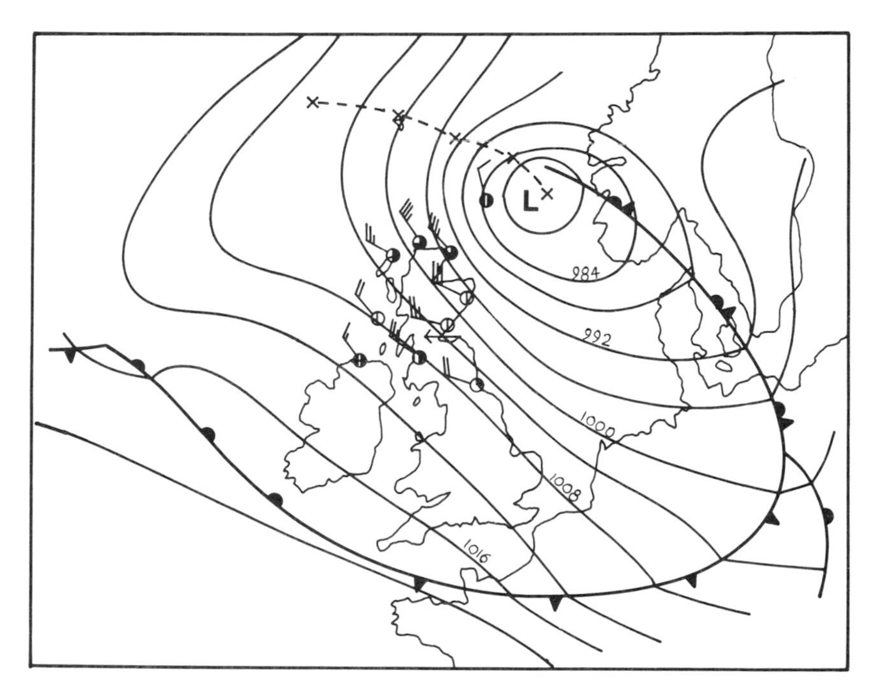

Fig. 44 Overland Kittiwake movements 15th November, 1973, 1200 hours. Large overland passage of Kittiwakes across central Scotland (arrowed) from North Sea gales to lighter winds off west coast. Track of depression marked (crosses) at six-hour intervals.

wind speed and sea state are the most important, and avoidance of strong winds leads to a greater abundance in cols and on the periphery of anticyclones, and in shallow depressions. In that respect, their behaviour parallels that of Fulmars.

In summer, Kittiwake movement off northeast Scotland is greatest during quiet anticyclonic weather, and doubtless comprises feeding birds from the vast local colonies (Table 5). When making feeding movements, Kittiwakes from Shetland colonies fly at a higher altitude with following winds than with opposing winds, except possibly when strong (over 10 m/s). The birds fly low in crosswinds to reduce drift. All these observations point to the fact that Kittiwakes are liable to displacement during gales, and indeed there is a higher percentage of large movements along coasts during onshore gales.

Skuas make similar visible coastal movements under such wind regimes. Autumn movements off east coasts of Britain often coincide with north-westerly gales, but many naturally move with the species which they parasitise. R. Furness[53] maintained that Great Skuas on the island of Foula intensified predation and kleptoparasitism when shoals of fish were scarce or difficult to catch in poor weather. Early in the season the skuas feed on dense fish shoals by plunge diving, and turn to predation (which is time and energy consuming) chiefly when shoals move into deeper waters later in the season. During summer, skuas are most abundant on the cliffs of Hermaness in Shetland during sea fog – a feature of this season when warm airmasses move over cooler seas. In clear weather, the skuas feed well out to sea. Kleptoparasitism is also an alternative feeding method of frigate-birds. These tropical seabirds are known to pluck fish from the wave crests and troughs of rough seas, but parasitise other birds in calm weather.[51]

Little else is on record regarding the prevention of feeding by fog, presumably due to the difficulty in making direct observations. There is evidence that terns bring less fish to their young in foggy weather, and chicks lose weight on such days. For kleptoparasitism the limiting visibility may be higher than that limiting direct feeding from the sea surface. Dense fog, with visibilities below 200 m, is uncommon at sea, but is more frequent in more northern, cooler waters, and over cold currents.

Wrecks

Although the mortality rate of adult seabirds by starvation is often difficult to assess, events occur from time to time which illustrate the hazards to which they are occasionally subject. It would seem that the smaller lighter seabirds, and those that feed from the surface rather than by underwater pursuit or diving, are more prone to prolonged periods of food scarcity. Gales themselves do little harm when displacing birds, provided that there is plenty of sea room. However, at times the protracted inability to feed, brought about by continuous gale-force winds, exhausts and weakens some species to the extent that they may then be blown long distances, also inshore and over land. Loss of weight ensues as fat reserves are used up, leaving birds without insulation from their cold wet environment. Spectacular 'wrecks' have been recorded on both sides of the North Atlantic, and it is the smaller species which succumb most often, particularly the Leach's Petrel and the Little Auk.

Small petrels frequently fly within a few centimetres of the sea surface, even in a heavy swell. They take advantage of the light airflow in the wave troughs and the lift found over the slopes. In gales, flight above the wave crests is prohibitive, since the birds would be blown quickly downwind. Problems arise especially in sudden shifts of wind, such as might occur at the passage of active fronts or depression centres, as it is then that swells become very confused and wave troughs less organised. The birds need to fly more vigorously, and since feeding becomes more difficult, exhaustion may overtake them after a while. The Wilson's Petrel appears to be relatively immune to wrecks – probably because it is adapted to an environment in the southern oceans where gales are very frequent and prolonged. Many other petrels have a subtropical distribution and are not likely to meet persistent gales, and the Storm Petrel also seems to be less affected than the Leach's Petrel. Peter Davis[30] recorded that the frequency of feeds brought to young Storm Petrels was little influenced by rough weather, although smaller fish

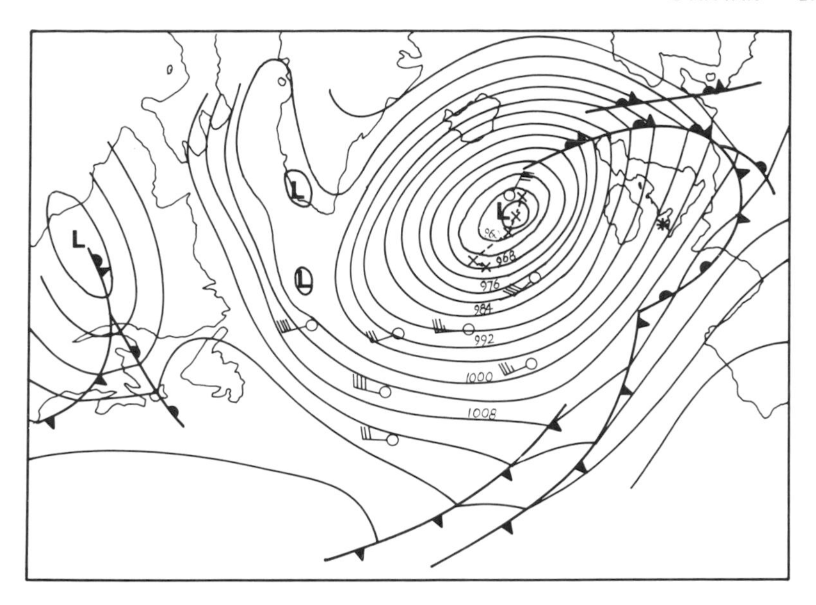

Fig. 45 23rd October, 1952, 1200 hours. Massive wreck of Leach's Petrels in southwest Britain and continental Europe resulted from prolonged gales in circulation of deep depression. The track of the low centre is shown marked at 12-hour intervals. The central pressure remained below 960 mbar for several days. *Bristol Channel.

may be more prevalent in the feeds.

The Leach's Petrel is subject to wrecks more than its relatives, and one of the most notable events on record is that of 1952. In late October, a deep stationary depression was centred about 800 km west of Scotland with prolonged winds of up to 32 m/s to the south of the centre on 22nd and 23rd (Fig. 45). It was not so much the strength of the winds as their persistence that caused feeding problems for enormous numbers of seabirds. Wave heights in the vicinity of the depression exceeded 8 m for long periods. With the depression being slow-moving, the very strong airflow gave trajectories that looped round the centre, so that the petrels probably spent a long time in continuously stormy conditions. Unable to feed efficiently, they became weak and emaciated, eventually drifting downwind on to the western shores of the British Isles where they were funnelled into the Bristol Channel. From the 21st October to 8th November, more than 6,700 dead or dying birds were reported in Britain and Ireland – a third of these in Bridgwater Bay, Somerset.[16] One sample of birds weighed 35% less than normal. Many were found in Europe – 70 in Belgium, several in France and even one in Switzerland. In autumn, Leach's Petrels linger in high latitudes, where they are most likely to meet storms. Additionally, numbers in the eastern Atlantic are swollen in October and November by North American birds. Wrecks do not always occur in such weather situations, however, and other seabirds are not always involved, so that feeding problems may not be the entire explanation. It has been suggested[25] that some unidentified epidemic disease is a contributory factor – one which perhaps operates in other wrecks of a single species.

A more recent coastal concentration of this species took place in mid September 1978, after a succession of deep depressions had moved ENE across the North Atlantic, culminating on 10th and 11th in a very strong westerly airstream, veering northwesterly. Further gales occurred during the remainder of the month, particularly on 16th when the remains of hurricane 'Flossie' moved across the Faeroes (Fig. 22). Thousands of petrels were blown into the Irish Sea and along the coast of northwest England and the eastern shores of the North Sea. A number on the Cheshire coast attempted to feed in beach pools, on wet sand and along the tide edge.

Some shearwaters are liable to wrecking and records of inland individuals are invariably preceded by severe westerly gales. The Fulmar – a species which, unlike others of its family, winters in the North Atlantic – has to tolerate much rougher weather. Fulmar wrecks, or at least periods of starvation, may be quite common in winter, although not often recorded. The species is less suited to dynamic soaring than its congeners in the southern oceans, and its greater dependence on the air currents deflected upwards by the waves may make it more prone to exhaustion or starvation in confused seas. Such a wreck occurred in early 1962. January and early February were characterised by severe westerly gales, more frequent and widespread than usual, with wave heights over 5 m particularly common in the western Atlantic. In addition to the curtailment of feeding by the very rough seas, black ice (frozen sea spray) affected those fishing vessels that remained at sea, reducing the Fulmars' additional food supply of fish offal. Weakened birds were carried into the Norwegian Sea in the circulations of two intense mid February depressions, and on 17th February, northerly gales in this area drove many birds south into the North Sea. During the period 22nd February to 15th March, well over 650 dead and emaciated Fulmars (together with Kittiwakes) were discovered along the east coasts of England, and further birds were drifted on to coasts in the eastern North Sea and Baltic. There were numerous examples of smaller, darker high latitude Fulmars, apparently including some of the small-billed Baffin Island population.

Of the auks, the Little Auk is subject to large wrecks most often. Fisher and Lockley,[51] in their classic book on seabirds, distinguished between 'flights', in which birds are drifted close enough to coasts to be observed in large numbers, and 'wrecks', where dead and dying birds are washed ashore or blown inland. They listed major flights and wrecks of the Little Auk on both sides of the North Atlantic in the first half of the present century, pointing out that they are wholly irregular and do not reflect population changes. Most wrecks, in fact, occur at the southernmost extremity of the winter range, and not within the normal wintering area. It appears, therefore, that the gales which cause the wrecks are affecting already weakened birds that have previously moved beyond their normal limits because of some unknown factor, probably food shortage.

Important wrecks of other auks occur,[6] with pollution and the stress of moult playing roles,[73] and individuals of most species are recorded inland from time to time. In February 1983, over 34,000 seabirds (mainly Razorbills and Guillemots from British colonies) were beached along the North Sea coast of Britain. These birds had probably sought shelter from severe westerly gales prevalent in January, only to be driven ashore weak from starvation by a spell of bitter NE gales during early February.[213] Despite the wreck, no significant reduction in numbers or breeding success was found in monitored colonies.

Skuas and gulls are not normally prone to wrecks, but not infrequently appear in large numbers in certain areas. Ringing recoveries show September and October to be the months of highest mortality of young Great Skuas. A wreck of Great Skuas in September, 1963 produced several recoveries inland in Europe, and an unusually high number of Shetland-ringed immature birds along the German and Dutch coasts. The gales, which were of hurricane-force in northern Britain on 26th September, concentrated large numbers of seabirds in the German Bight area during the first three weeks of October (Fig. 27).

A few records exist of small wrecks of Long-tailed Skuas. Although they are smaller and lighter than other skuas, wrecks may depend only on prolonged disturbances of feeding on their normal migration route, thought to be in mid Atlantic. They are, however, more common in autumn in the North Sea than on the west coasts of Britain, perhaps indicating a displacement to calmer, more sheltered, waters to feed. If gales in this area then interrupt feeding, birds may become weak enough to be storm-driven.

Other species, not strictly seabirds but wintering at sea, are occasionally storm-driven inland and into coastal waters. Red-throated Divers appear at times, but two phalaropes which winter at sea, the Grey and Red-necked, are more frequent. Like the small petrels, they are light and unable to remain waterborne in gales. The former species has occasionally been recorded in large numbers inshore while on passage on both sides of the North Atlantic, and in Newfoundland these concentrations after prolonged gales have earned it the name 'gale-bird'.

Although in lower latitudes seabirds are relatively free of extensive strong wind systems, they occasionally come under the influence of tropical storms. In the North Atlantic these are most frequent from July to October, sometimes curving into the mid latitude westerlies to become extra-tropical depressions (Chapters 1 and 8). In the North Atlantic, a storm is technically known as a hurricane when sustained winds reach 33 m/s or more – between 17 m/s and 33 m/s it is merely called a tropical storm. They have their counterparts in other oceans – typhoons, cyclones, and willy-willys are all storms with similar characteristics. The area covered by the storm is very small compared to that of a depression. In the centre of an intense vortex, warm subsiding air creates a relatively calm, clear, 'eye', normally some 20 to 40 km in diameter. The eye is surrounded by a wall of very strong upcurrents and massive Cumulonimbus cloud, within which are violent low level winds blowing cyclonically. Storms are notoriously fickle in their movement, and prediction of future tracks is very difficult.

Oceanic birds probably do not attempt to avoid hurricanes – merely remaining in the wind system and drifting when they find it too strong to fly against. If they reach the eye, they may well rest on the water – perhaps moving along in the light winds and confused seas. Whatever their position in relation to the eye, there is little doubt that they can be carried great distances outside their normal range. Vagrant subtropical and other seabirds are found well north along the American east coast and occasionally considerable distances inland. In October 1968, hurricane 'Gladys' (Atlantic hurricanes used all to be given female names until recently – now male and female names are used alternately, in alphabetical order) deposited tens of thousands of Laughing Gulls and around a thousand Black Skimmers in Nova Scotia, having carried them as much as 2,000 km from their normal range (Fig. 46). The birds had probably been transported within the storm centre, especially the weaker flying skimmers. Although there were reports of dying birds, the majority of these storm-driven waifs were thought to have returned south at a later date. Certainly, skimmers were observed along the coasts further south, but the gulls apparently dispersed southwards at sea.[106] A similar event happened in 1958, when hurricane 'Helene' carried these species as far northeast as Newfoundland. As in all storms, the resultant distribution of seabirds is largely dependent upon the phase of their annual cycle with regard to migratory and breeding states, and previous weather conditions.

BREEDING

Oceanic seabirds come ashore only by accident or to breed. They are colonial – often breeding in vast numbers in very exposed places. Their arrival at breeding sites from oceanic wanderings varies with latitude and climate, for in polar waters the weather early in the season is too inimical for breeding to begin. In temperate regions, some remain at or near their colonies in winter, departing to sea for only short periods at a time. Such individuals, especially of those species which can be found far out to sea in winter, comprise only a small part of the population. Their presence ashore is often partly controlled by the weather.

From November, Fulmars at British colonies are fewer when there have been strong winds for the previous few days. When winds freshen, many leave the cliffs, since these conditions provide them with suitable assistance to fly to feeding areas. In January and February, the number of birds at the colony can be interpreted in terms of the wind conditions over the past three days. In January, birds remain on ledges in calm weather, even when snow

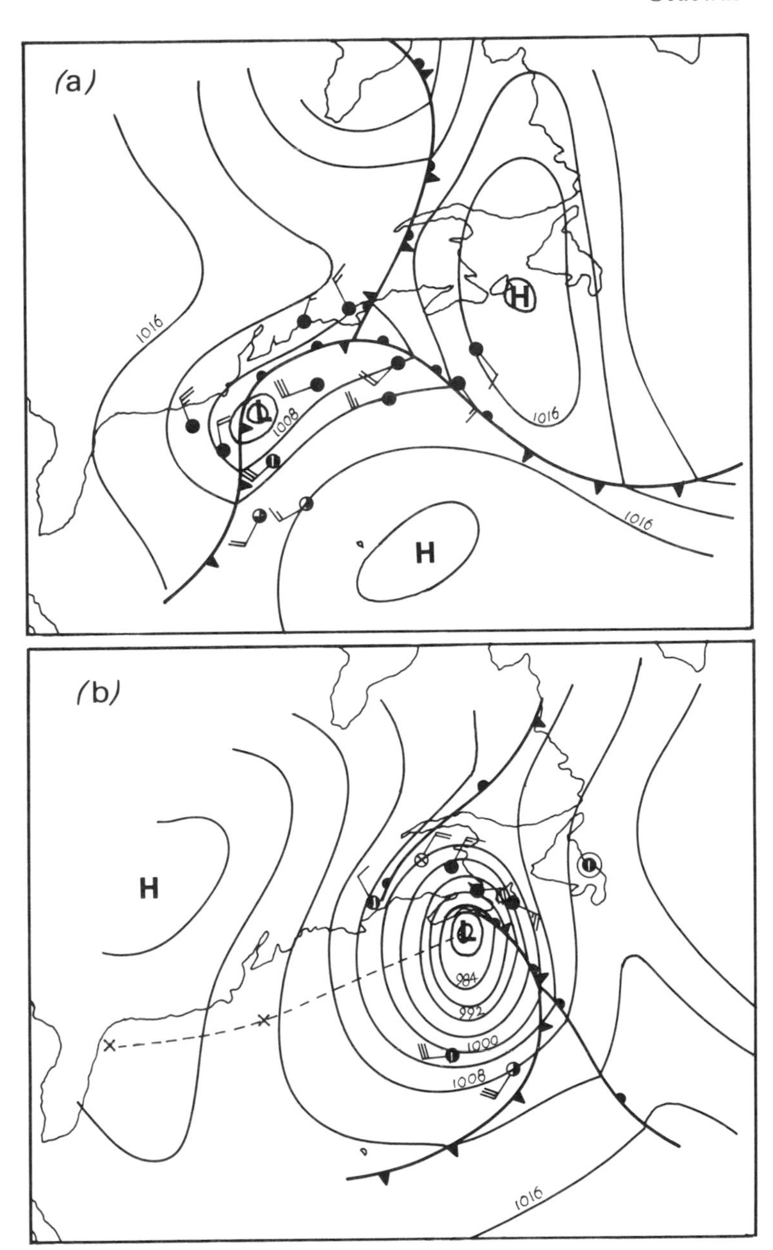

Fig. 46 (a) 20th October, 1968, 1200 hours.
(b) 21st October, 1968, 1200 hours.

L eye of hurricane 'Gladys'

Hurricane 'Gladys' off Cape Hatteras on 20th October became extratropical and deepened rapidly. It moved NE, carrying with it large numbers of seabirds into Nova Scotia, subsequently filling and drifting slowly east off Newfoundland. Track of centre marked at 24-hourly intervals.

and ice are present and temperatures are well below zero, but precipitation, combined with other deteriorations in the weather, can reduce attendance.[35] In summer, there is no apparent correlation between attendance and any weather element – be it wind speed or direction, temperature, cloud cover or fog – although in April, massive departures have occasionally been observed during violent onshore winds. Onshore gales and breaking seas may also drive Fulmars from the more exposed ledges.

In many temperate zone colonies, Fulmars are only absent completely in September, returning in some numbers in October during good weather. Auks similarly appear at this time, and such winter visits have become more frequent and earlier in northern Britain. However, these changes are thought to be related more to population dynamics than to weather conditions, and there is often little correlation among species. Before laying, and occasionally afterwards, there appear to be declines at Guillemot colonies in strong winds and rain. Birkhead recorded a mean of 20–30% of off-duty breeding birds to be present at the Skomer colony with winds less than force 5 (9 m/s).[12] Only above this strength was there a significant drop in the numbers present. No significant relationship has been found between the weather and the presence of Razorbills[140] or Puffins[205] during breeding, although gales do seem to upset attendance patterns early in the season.[25]

Seabird breeding both in temperate and polar regions is timed to coincide with the summer blooming of plankton and the corresponding increase in that food source's marine predators. The average date of laying is more constant than in passerines due to smaller variations in the food supply, except in the arctic where the break-up of sea ice varies from year to year. A late thaw prevents seabirds from reaching feeding grounds within their operational range, and successful breeding depends upon the ability to find food within easy reach of colonies. In Canada, Brunnich's Guillemots do not leave the colony in stormy weather. In the incubation period, the guillemots may forage up to 200 km to visit the edge of the fast ice where Arctic Cod concentrate. This rich food supply, used also by Fulmars and Black Guillemots, is found where low salinity combines with warmer water in freshwater outlets. The positions of the most crucial feeding sites may alter substantially from year to year due to changes in the ice edge – changes which determine the summer distribution of breeding birds. These variations so affect food abundance and accessibility that the chief feeding sites can alter in position over short periods of time. As we saw in Chapter 5, ice-melt depends on the track and intensity of arctic depressions, and sea ice distribution alters rapidly with changing winds. Only slight sea surface temperature anomalies are necessary to delay the break-up of ice in the high arctic, especially in narrow sounds between islands, so that feeding in open leads is impossible for returning seabirds. Anomalously low water temperatures and winds from directions which hold back warmer water currents disrupt the breeding cycle.

For long distance migrants returning from the southern hemisphere or the

Kittiwakes brooding at base of cliff

subtropics, weather on passage can influence arrival times. Kenneth Williamson[188] reasoned that, if delays occur, adult birds (which in several species are first to arrive on the breeding grounds) may be overtaken by younger individuals. He found that Arctic Skua divorces on Fair Isle were common when one of a pair was delayed, in which case the other then often mated with an unattached bird.

There is no direct correlation between the weather and the return of the Gannet at the Bass Rock,[113] but the variation in the date of return is probably the result of local fishing conditions. In extremely stormy winters, their arrival may be delayed, but never beyond late February. In the Gulf of St Lawrence, however, snow and ice at the colonies delays their arrival until as late as May. This makes the breeding cycle of the Gannet, and indeed other seabirds, more compressed in polar climates, with a rapid start to laying immediately following their arrival.[114] A similar situation is found in arctic land birds, of course, but for them food is generally much scarcer at the beginning of the season.

Even in temperate climates, the physical conditions at the nest site have been found to delay laying. For example, in 1975 waterlogging of parts of the Farne Islands Puffin colony delayed laying until the soil dried out. Above average spring rainfall waterlogged the soil to within the burrowing zone (the top 20–25 cm) and the eventual distribution of eggs and chicks reflected the variation in soil conditions. However, considerable physical problems do not deter Manx Shearwaters from occupying burrows in their mountain-top colonies on the Isle of Rhum. Here, at an altitude of over 500 m, they return in the third week of March in extremes of wind, rain, cloud and humidity. The birds often need to clear away drifted snow to gain access to the burrows, and at times, even in April, the soil is frozen to a depth of several centimetres.

The exposed environments of seabird nests afford little protection from the elements except for those in burrows. In severe gales there can be a high egg loss when the violence of the wind and updraughts at cliffs effectively clear nests on flat tops and ledges. Even sitting Gannets have been blown from their nests. Low level Kittiwake nests are regularly destroyed in summer storms, and wave action is a problem for all seabirds nesting near cliff bases and in caves.

In June, 1973, an intense depression brought easterly winds of 17–21 m/s to seabird colonies in Newfoundland. Only a small percentage of the Herring and Ring-billed Gull chick population survived the storm, while Kittiwake nests up to 12 m above the sea, and also Guillemot and Razorbill eggs, were washed away by high seas. In higher exposed areas, many young were killed by rain and low temperatures, particularly where heavy rain had created torrents which washed nest contents away. After the storm both Herring Gulls and Kittiwakes formed large flocks consisting of failed breeders which (because of the lateness of the season) were unable to re-lay, though both auks had re-laid by early July.[173]

Terns are particularly prone to dramatic breeding failures in poor weather. Strong winds raise sand and cover nests, eggs and young, while winds combined with high tides wash away whole colonies, though Dunn[40] found that in Northumberland, Roseate chicks in burrows were rather less prone to chilling than other tern chicks above ground. A certain amount of drifting sand can be tolerated by a nesting tern; it can scratch away accumulations or protect its young by careful brooding. Common Terns frequently nest on 'windrow', the jetsam of dead vegetation just above high water mark. In New Jersey, USA, a high proportion of these nests survive tidal flooding, which is most severe when spring tides coincide with high winds. The nests on windrow float at high tide and resettle when the sea recedes.

As mentioned earlier, wind speed and the associated sea state alter fishing success and have a marked effect on the growth rates of young terns. Dunn[40] found that in Northumberland Roseate Terns are the most sensitive of the local breeding terns, and calculated that a wind of 5 m/s depressed the weight increase of chicks by two-thirds, possibly because in Britain the species is at its northern limit and thus less well adapted to prevailing meteorological conditions. Growth of Arctic Tern chicks, on the other hand, is not retarded by strong winds, for this is a well-adapted pelagic northern species. Like other colonial seabirds whose nests are often close together (e.g. Gannets), the posture of incubating terns normally depends on social factors, but as wind speed increases, these factors are overridden so that all birds face the wind when it is strong.[59]

Some subtropical terns attempt to cool eggs and young with moisture in hot weather. This behaviour has also been recorded in the Netherlands, where R. Mes and others[105] observed that Common Terns changed their activity pattern on hot days. Whereas birds arriving at a colony with prey did so mostly in the cooler hours, those without prey were more active in the middle of hot days when the temperature exceeded 24°C. Those birds, after skimming the water and wetting the underparts and possibly the brood patch, returned immediately to the nest. Mes and his colleagues proposed that the terns used the moisture on the plumage to cool eggs or young on the hot sand. Recent observations in the USA indicate that such cooling is ineffective for older, larger gull and tern chicks, which suffer high mortality when air temperatures exceed 31°C.[209]

In regions subject to tropical storms, such as the eastern seaboard of the USA, seabird colonies are at risk. Although these storms are most frequent after the breeding season has finished, the occasional early hurricane causes havoc, not only with its high winds but also with the rough seas and exceptionally high tides that it generates. The very low pressure in the eye of the storm raises the sea level (the 'hurricane wave') so that its passage along coastal areas creates serious flooding. In June, 1966, hurricane 'Alma' decimated Sooty Tern colonies on the Florida islands by burying many chicks alive in the sand and killing others. Hurricane 'Agnes' in June, 1972 also caused destruction as it moved north into Florida. A few days later it passed east of New Jersey with renewed activity, where incubating Common Terns were observed to sit firm in gusts of 25 m/s, occasionally readjusting plumage and eggs as the sand piled against the breast feathers. In some nests the eggs were 7 cm higher after the storm than before.[59] As indicated by the initial letter of these named June storms, they are invariably the first of the season.

Hurricane 'Agnes' formed in the Caribbean from clouds originating south of the equator in an event which itself causes immense seabird mortality by starvation and breeding failure. The cold Peruvian current off the west coast of South America, where the southeast Trade winds move surface warm water away to allow cold upwelling, is inhabited by millions of Guanay Cormorants, Peruvian Boobies and Brown Pelicans, which feed on the vast multitudes of fish. Every few years, a weakening of the Trades allows warmer surface water nearer the coast, restricting the upwelling, killing the fish and reducing the avian populations. The northerly current which accompanies this is known as El Niño (the child), since it is prevalent around Christmas time. The warm water gives rise to enhanced convection, thus resulting in a displacement of the normal vertical atmospheric circulation in the tropics. Indeed, El Niño events, particularly marked in 1972 and 1982, are interrelated to the Sahel drought and worldwide alterations in monsoon and tropical storm activity. The effects of the 1982 event were prolonged and widespread throughout low latitudes in the Pacific Ocean, with catastrophic breeding failures and redistributions of island and coastal bird populations. Similar, but less severe, related phenomena have occurred in both the South Atlantic and Indian Oceans.

In Chapter 5, I discussed the adaptation of young to periods of starvation, particularly in the Swift. The young of some pelagic seabirds (e.g. Storm

Petrel and Manx Shearwater) show similar adaptations. Changes in feeding conditions due to weather affect these nestlings only temporarily and, indeed, variations in feeding rhythms occur even without food shortages. Nevertheless, if the adults have difficulty in feeding in rough seas, their chicks may pass up to two to three days without food, developing slowly at a slightly low body temperature to reduce the metabolism.

Certain shearwaters and petrels return at night to feed the young or relieve a mate at the nest. This helps to reduce predation by gulls. Off Welsh breeding islands, Manx Shearwaters gather on the sea at dusk. Flocks are close inshore in heavy seas and strong winds, but in a flat calm remain 1·5 to 3 km offshore. Less searoom is necessary in rough weather owing to the easier manoeuvrability. Cloud cover and the state of the moon appear to dictate both the degree and timing of predation. Additionally, unattached and sub-adult non-breeders show different behaviour to breeding birds. When the moon is bright non-breeding birds tend not to visit breeding sites, and those that do come ashore call less often. On the contrary, the frequency of visits of breeding adults is not affected by the state of the moon or the weather.

On clear moonlit nights, some breeding Storm Petrels fly directly to and from burrows, while others display.[25] On dark nights, incoming or outgoing shearwaters on the Welsh island of Skomer blunder into gull nesting areas, and if misty collide with rocks and other objects. Predation of Manx Shearwaters by Great Black-backed Gulls is low on clear moonlit nights, but high after moonless or dark foggy nights when the number of visiting non-breeders increases markedly. Gulls are able to attack birds earlier on moonlit nights than when overcast – predation is then more common in the morning.[25] Although the threat of predation may be an important factor in the evolution of nocturnal visits, it appears to have no significance in population regulation. Inter-colony variation in nocturnal behaviour may result from local densities of predators.

Olfactory navigation probably plays a leading role in the location of nests during these nocturnal visits. Some Leach's Petrels in New Brunswick, Canada, nest in burrows beneath dense conifer forest. Grubb[64] found that the method of arrival of an individual varies with wind conditions, indicating location and guidance by the odour emanating from its burrow. Petrels land closer to burrows in still air than in wind, and approach randomly. When there is a wind blowing, the birds drop to the forest floor downwind from the burrows and walk upwind.

Young petrels and shearwaters fledge at night, also to avoid predators. R. M. Lockley[92] watched fledging Manx Shearwaters on Skokholm making their way always downhill from burrows to the sea in calm weather, when they had no assistance from the wind. In a fresh wind, they headed into the wind, even uphill, and were able to gain enough lift to make their first flight, which eventually led them to the windward coast and the sea.

Even if the young birds reach the sea successfully, their problems are not over. With onshore gales in late summer, newly-fledged young are at risk, being caught inshore and often drowned in the breakers. Young Fulmars fledge at weights much higher than adults, and are particularly prone to beaching. The often violent and difficult eddies at the cliffs can toss them into clifftop fields where they are unable to become airborne again owing to their weight. Ornithologists in Fulmar breeding areas in late summer are familiar with oil-spitting youngsters marooned a little way inland. The mainland coast opposite the islands off South Wales occasionally receives thousands of dead young Manx Shearwaters and Gannets when severe onshore winds occur in autumn.

Like Fulmars, fledgling Gannets have problems in the violent updraughts around cliffs. Bryan Nelson[114] watched a bird turned completely round by the wind, whereupon it plummeted 75 m on to rocks. He described vividly the behaviour of the young Gannet before and after fledging. At the onset of rain, the young are stimulated to flap their wings – a myriad of black wings raised and flapped while the adults remain unmoved. He thought this to be a tactile stimulation which releases bathing-flapping, the first act upon landing on the sea. When fledging, the young bird will soar immediately if a fresh onshore wind is blowing. The fat reserve, like that of petrels and shearwaters, keeps it alive for one to two weeks, dependent upon wind and temperature, while it learns to fish. The weather as a factor in survival becomes more critical as the fat is used up. A young Gannet can become airborne more easily when light, but equally at this stage it is more imperative to feed, so that a spell of poor weather and rough seas may prove fatal. For that reason, early fledgers are at an advantage over those fledging late, since the latter run a greater risk from the more frequent stormy periods in autumn.

MIGRATION

The broad principles outlined in the chapters on migration apply also to seabirds. However, many aspects of seabird migration are poorly understood, since movements are so interwoven with feeding behaviour that interpretation is often difficult. Seabird migration has some affinity with that of land birds. Northward movement in spring is often undertaken in fine anticyclonic weather with southerly winds, while southward passage in autumn is frequently associated with polar highs in the clearing weather to the rear of depressions. On meeting poor conditions, several seabird species settle on the sea, particularly the inefficient fliers, such as auks. Seabirds naturally experience less difficulty in poor weather than land birds, since they are always in their natural environment, despite the possibility of food scarcity when migrating over barren stretches of ocean.

Some pelagic species such as shearwaters and albatrosses may make use of planetary wind systems, migrating with the low level circulation round subtropical anticyclones, and in mid latitude westerlies. Indeed, altitude of most transoceanic seabird migration is low. The more immediate effects of the weather are complex. Needless to say, navigation can be as precise as in other avian groups, and the probable use of astronomical cues has been demonstrated in the cases of the Manx Shearwater. The now famous experiments performed with this species in the middle decades of this century involved birds released in both fine weather and overcast conditions, having been removed great distances from Welsh breeding colonies. Birds homed very rapidly to their colony from as far afield as USA, Italy and Switzerland, with one bird flying over 5,000km from Boston, USA to Skokholm in just 12½ days. Those released in overcast weather were unable to take up their correct heading and scattered randomly from the release point, or drifted downwind, their homing ability apparently impaired. In clear skies inland birds set out in directions indicative of direct flight overland, rather than a roundabout sea track.[51,92]

Studies of Manx Shearwater passage off western Ireland (at a headland which protrudes into the southward migration route and therefore gives a better indication of passage rather than feeding movements) showed that autumn migration takes place in anticyclonic conditions or in rising pressure and northwesterly winds during the clearing weather to the rear of a depression.[138] Few birds move in disturbed weather, and little migration begins under overcast skies. If passage is in force, it decreases when a slow moving depression crosses the route. The briefer period of poor weather in a fast moving depression has less effect. Drift (and consequently greater passage observed from the coast) is observed in onshore winds, but in poor visibility the shearwaters seem not to hug the coast as they do in movements at other sites such as Cape Clear. In headwinds, the birds reduce speed and fly low over the waves in the lighter windflow.

Gannet migration is similarly complex. There is a prolonged departure from colonies in autumn, with occasional mass exits during fine weather. There appears to be a considerable variation in the effect of wind on passage, and juveniles tend to be drifted inshore more than adults. Mass inshore movements occur together with other seabirds.[114] Off northeast Scotland, the passage of Gannets in spring and autumn is higher in poor visibility associated with anticyclonic weather. Quiet conditions also stimulate the spring return and autumn departure of Kittiwakes, and terns, the larger gulls and the Cormorant invariably set out on migration in fine weather (Table 5). In spring, Common Gulls return to Scandinavia in following winds, particularly in fine weather behind an eastward moving front.[177] Radar observations in northeast Scotland suggest that birds overtaking a front descend until conditions have improved. Some must penetrate the frontal weather, since flocks occur in the northern isles in the poor visibility associated with pre-frontal east to southeast winds. Overland, spring passage of the Lesser Black-backed Gull is delayed by gales and thick cloud, during which the birds settle and make no attempt to fly into the wind.

Like shearwaters, migrant Pomarine Skuas tend to appear inshore during drift conditions. They move quickly northwards in spring, in fine weather with light south to east winds. Off southwest Ireland, however, most spring movements of this species are associated with stronger winds from a southerly point ahead of a front – indicative of onshore drift. Further north in the Hebrides, the largest spring movements (with other skuas in a northward direction) take place with winds between SSW and NW to the rear of fronts, or in the decreasing winds behind a depression. The former also indicates drift, and most movements are in winds of force 3 to 5 (5–10 m/s). Headwinds of up to force 5 are tolerated, but migration ceases in stronger headwinds.[29]

Outside the breeding season, the Little Gull is rather more oceanic in its habits than most gulls. Movements are complex, with a limited initial dispersal to allow moult in relatively sheltered areas. Later dispersal to sea occasionally leads to long distance movement across the Atlantic. Numbers of birds are recorded in the USA and Canada, where limited colonisation has taken place in the past 20 years. Two routes have been suggested. First, wintering gulls in southwest Europe come under the influence of northeast winds which lead into the Trade wind zone. Trade winds are extremely constant – probably the most constant of all the earth's wind systems – and we have already seen that a number of European birds have appeared in the New World using this route. The second route is possibly a northern one, though less likely. Dispersal to the north of Britain puts birds into wind flows that are frequently easterly to the north of depressions. A family of depressions in a mobile synoptic situation does indeed result in a generally easterly flow in the Atlantic between the Iceland/Faeroes region and eastern Canada.

Airflows on the northern and western flanks of these depressions assist the autumn and winter movements of several arctic species. The final stages of N. Atlantic depressions result in a slowmoving cyclonic circulation in the northeastern part of the ocean – the 'Icelandic' low is an aggregate of these (Figs. 47–50). Little Auks and Fulmars disperse to Greenland and Newfoundland waters,[25] while Ross's Gulls wander towards Britain. During the winters of 1980/81 and 1982/83, an increase in appearances of this high arctic vagrant resulted from frequent outbreaks of very strong north and northwest winds associated with a much intensified and displaced Icelandic low.[202]

Much of the autumn marsh tern passage in Britain (particularly Black Terns) is observed under drift conditions while the birds are migrating to wintering grounds in southwest Europe and northwest Africa. Their appearance in southern England follows north to east winds and poor weather ahead of approaching fronts. Autumn movements of Black and White-winged Black Terns suggestive of reversed passage have also taken place in anticyclonic weather. The return of the Black Tern in spring is initiated by rising temperatures. Large passages in Britain occur with a depression west of Spain, and an anticyclone over the North Sea or to the north of Britain. Northeast to east winds between the two systems deflect returning birds westwards. During a short mild spell in early May, 1974, when freshening east winds ahead of a Biscay depression drifted birds inshore along the English Channel, heavy coastal passage was observed, not only of Black Terns, but also of Common Terns, Arctic Skuas and Little Gulls. A parallel synoptic situation was present in May 1970. On this occasion, large numbers of Black Terns began to pass along the coast of south and east England, but were accompanied by White-winged Black and Whiskered Terns. This suggested an element of overshooting in warm weather induced by an anticyclonic spell immediately prior to this period, at a time noted for southern migrants.

The sea terns are all long distance migrants, one of them, the Arctic Tern, flying vast distances from the Arctic to the Antarctic. The main passage across the Antarctic Ocean has been admirably appraised by Salomonsen.[151] The terns are carried towards the wintering area at the mercy of the southern ocean storms. Since arrival must be achieved before wing moult begins at the end of December, late birds remain in African waters. Birds carried eastwards and occasionally found in southern Australia are probably caught up in fast moving waves on the southern polar front, rather than in the developing depressions. Salomonsen correlated movement from South Africa to the winter feeding area along the ice edge with trajectories using the northwesterlies on the north side of southeastward moving depressions. (Remember that depressions in the southern hemisphere move generally in an easterly direction as they do in the north, but have a clockwise rotation i.e. in the opposite sense. Thus, to compare directions with those in northern hemisphere circulations, only north and south components must be reversed. A southwest wind would therefore equate to a northwest wind in the northern hemisphere).

Once at the ice edge, the terns concentrate in the rich feeding grounds where upwelling takes place. Being still in the prevailing west wind zone on arrival, some (particularly young birds) may be carried east while in moult. The young birds moult later than adults, and inexperience tends to make them more prone to eastwards drift. Some birds remain during their first summer, and in the first two years of life may circumnavigate the globe. On departure at the end of the southern summer, the ice edge has withdrawn into the easterlies associated with the antarctic anticyclone, so that the terns move west and finally northeast towards South Africa on the western flank of the depressions. Other populations migrate up the west coast of South America.

SUMMARY

Seabirds are affected by the weather in rather different ways than are other groups of birds, owing to their peculiar habits and dependence on the sea. In winter they have either adapted to a successful existence in stormy oceans, or evolved a migration to less weather-affected seas. Migratory movements have some affinity to those of other groups of migrants, but most movements are complex, and because of the problems of observation, analysis and conclusions are difficult to achieve. Like landbirds, the weather influences both prey and the act of predation. In general, wind is by far the most important element influencing feeding. Prolonged gales, with associated rough seas, weaken some of the smaller species by preventing feeding. They may be driven great distances in a weakened state, and very occasionally undergo massive mortality.

The exposure of breeding sites varies among species, and is of importance to those nesting above ground and near the water's edge. The laying dates of many seabirds are very much less variable than those of landbirds, since food resources show only small fluctuations in the short-term, although polar seabirds are influenced by the break-up of sea ice.

It is clear, however, that most seabirds can easily tolerate weather conditions which man may find appalling. Perhaps it is this harmony with a rigorous environment, as well as the remote places in which they nest, that endears them to so many of us.

Selected bibliography

The following list is by no means complete, and a further more comprehensive list is available from the author. All references specifically mentioned in the text and in the Figures are listed, but many general references have necessarily been omitted.

1. ABLE, K. P. 1973. The role of weather variables and flight direction in determining the magnitude of nocturnal bird migration. *Ecology* 54: 1031–1041.
2. ALERSTAM, T. 1975. Crane *Grus grus* migration over sea and land. *Ibis* 117: 489–495.
3. —— 1976. Nocturnal migration of thrushes (*Turdus* spp.) in southern Sweden. *Oikos* 27: 457–475.
4. —— and ULFSTRAND, S. 1974. A radar study of the autumn migration of Woodpigeons *Columba palumbus* in southern Scandinavia. *Ibis* 116: 522–542.
5. ARMSTRONG, E. A. 1955. *The Wren*. London.
6. BAILEY, E. P. and DAVENPORT, G. H. 1972. Die-off of Common Murres on the Alaska peninsula and Unimak island. *Condor* 74: 215–219.
7. BAKER, K. 1977. Westward vagrancy of Siberian passerines in autumn 1975. *Bird Study* 24: 233–242.
8. BARLEE, J. 1957. The soaring of Gannets and other birds in standing waves. *Ibis* 99: 686–687.

9. BETTS, F. N. 1948. The flight of storks on migration. *Ibis* 90: 150–151.
10. BIBBY, C. J. 1981. Wintering Bitterns in Britain. *Brit. Birds* 74: 1–10.
11. BIRKHEAD, T. R. 1976. Effects of sea conditions on rates at which Guillemots feed chicks. *Brit. Birds* 69: 490–492.
12. —— 1978. Attendance patterns of Guillemot *Uria aalge* at breeding colonies on Skomer Island. *Ibis* 120: 219–229.
13. BLOKPOEL, H. and RICHARDSON, W. J. 1978. Weather and spring migration of snow geese across southern Manitoba. *Oikos* 30: 350–363.
14. BOURNE, W. R. P. 1973. Influx of Great Shearwaters in autumn 1973. *Brit. Birds* 66: 540.
15. —— 1982. The manner in which wind drift leads to seabird movements along the east coast of Scotland. *Ibis* 124: 81–88.
16. BOYD, H. 1954. The 'wreck' of Leach's Petrels in the autumn of 1952. *Brit. Birds* 47: 137–163.
17. —— 1957. Mortality and fertility of the White-fronted Goose. *Bird Study* 4: 80–93.
18. BRADBURY, D. L. and PALMÉN, E. 1953. On the existence of a polar-front zone at the 500-mb level. *Bull. Am. Meteorol. Soc.* 34: 56–62.
19. BRODIE, J. 1976. The flight behaviour of Starlings at a winter roost. *Brit. Birds* 69: 51–60.
20. BROWN, L. H. 1976. *British Birds of Prey*. London.
21. BUXTON, J. 1950. *The Redstart*. London.
22. CALDER, W. A. and KING, J. R. 1974. Thermal and caloric relations of birds. In FARNER, D. S. and KING, J. R. (eds) *Avian Biology*, Vol. 4, New York.
23. CHANDLER, R. J. Influxes into Britain and Ireland of Red-necked Grebes and other waterbirds during winter 1978/79. *Brit. Birds* 74: 55–81.
24. CONE, C. D. JR. 1968. Thermal soaring by Starlings. *Auk* 85: 19–23.
25. CRAMP, S., BOURNE, W. R. P. and SAUNDERS, D. 1974. *The Seabirds of Britain and Ireland*. London.
26. CRAMP, S. *et al.* 1977–88. *The Birds of the Western Palearctic*. Vols. I–V. Oxford.
27. CRISP, D. J. (Ed.) 1964. The effects of the severe winter of 1962–63 on marine life in Britain. *J. Anim. Ecol.* 33: 165–210.
28. DARE, P. J. and MERCER, A. J. 1973. Foods of the Oystercatcher in Morecambe Bay, Lancashire. *Bird Study* 20: 173–184.
29. DAVENPORT, D. L. 1979. Spring passage of skuas at Balranald, North Uist. *Scott. Birds* 10: 216–221.
30. DAVIS, P. 1957. The breeding of the Storm Petrel. *Brit. Birds* 50: 85–101; 371–384.
31. —— 1966. The great immigration of early September 1965. *Brit. Birds* 59: 353–376.
32. DENNIS, R. H. 1970. Spring migration. *Fair Isle Bird Obs. Rep.* No. 22 (1969): 11–17.
33. DIERSCHKE, F. 1976. Auswirkungen der Sturmschäden vom 13.11.1972 auf die Sommervogelbestände in Kiefernforsten der Lüneburger Heide. *Die Vogelwelt* 97: 1–15.
34. DOBINSON, H. M. and RICHARDS, A. J. 1964. The effects of the severe winter of 1962/63 on birds in Britain. *Brit. Birds* 57: 373–434.
35. DOTT, H. E. M. 1975. Fulmars at colonies: time of day and weather. *Bird Study* 22: 255–259.
36. DUGAN, P. J., EVANS, P. R., GOODYER, L. R. and DAVIDSON, N. C. 1981. Winter fat reserves in shorebirds: disturbance of regulated levels by severe weather conditions. *Ibis* 123: 359–363.

37. DURAND, A. L. 1963. A remarkable fall of American land-birds on the 'Mauretania', New York to Southampton, October 1962. *Brit. Birds* 56: 157–164.
38. —— 1972. Landbirds over the North Atlantic: unpublished records 1961–65 and thoughts a decade later. *Brit. Birds* 65: 428–442.
39. DUNN, E. K. 1973. Changes in fishing ability of terns associated with windspeed and sea surface conditions. *Nature* 244: 520–521.
40. —— 1975. The rôle of environmental factors in the growth of tern chicks. *J. Anim. Ecol.* 44: 743–754.
41. EAST, M. 1980. Sex differences and the effect of temperature on the foraging behaviour of Robins *Erithacus rubecula*. *Ibis* 122: 517–520.
42. ELKINS, N. 1965. The effect of weather on the Long-tailed Duck in Lewis. *Bird Study* 12: 132–134.
43. —— 1979a. High altitude flight by swans. *Brit. Birds* 72: 238–239.
44. —— 1979b. Nearctic landbirds in Britain and Ireland – a meteorological analysis. *Brit. Birds* 72: 417–433.
45. —— and ETHERIDGE, B. 1974. The Crag Martin in winter quarters at Gibraltar. *Brit. Birds* 67: 376–387.
46. —— and WILLIAMS, M. R. 1972. Aspects of seabird movement off northeast Scotland. *Scott. Birds* 7: 66–75.
47. ENGLAND, M. D. 1966. Great Bustards in Portugal. *Brit. Birds* 59: 22–27.
48. EVANS, P. R. 1968. Autumn movements and orientation of waders in northeast England and southern Scotland, studied by radar. *Bird Study* 15: 53–64.
49. FEARE, C. J. 1966. The winter feeding of the Purple Sandpiper. *Brit. Birds* 59: 165–179.
50. FINLAYSON, J. C., GARCIA, E. F. J., MOSQUERA, M. A. and BOURNE, W. R. P. 1976. Raptor migration across the Strait of Gibraltar. *Brit. Birds* 69: 77–87.
51. FISHER, J. and LOCKLEY, R. M. 1954. *Sea-Birds*. London.
52. FREETHY, R. 1980. Moorhens' rapid construction of brood nest. *Brit. Birds* 73: 35.
53. FURNESS, R. W. 1977. Great Skuas as predators of mammals. *Scott. Birds* 9: 319–321.
54. GEIGER, R. 1965. *The Climate Near the Ground*. Cambridge, Mass.
55. GIBB, J. 1960. Populations of tits and Goldcrests and their food supply in pine plantations. *Ibis* 102: 163–208.
56. GLADWIN, T. W. and NAU, B. S. 1964. A study of Swift weights. *Brit. Birds* 57: 344–356.
57. GLUE, D. 1972. Observations on the passage of migrant geese and waders in northwest Iceland. *Bird Study* 19: 252–254.
58. GLUE, D. E. and NUTTALL, J. 1971. Adverse climatic conditions affecting the diet of a Barn Owl in Lancashire. *Bird Study* 18: 33–34.
59. GOCHFELD, M. 1978. Incubation behaviour in Common Terns: influence of wind speed and direction on orientation of incubating adults. *Anim. Behav.* 26: 848–851.
60. GORDON, S. 1955. *The Golden Eagle, King of Birds*. London.
61. GOSS-CUSTARD, J. D. 1969. The winter feeding ecology of the Redshank *Tringa totanus*. *Ibis* 111: 338–356.
62. —— 1976. Variation in the dispersion of Redshank *Tringa totanus* on their winter feeding grounds. *Ibis* 118: 257–263.
63. GREEN, G. H., GREENWOOD, J. J. D. and LLOYD, C. S. 1977. The influence of snow conditions on the date of breeding of wading birds in north-east Greenland. *J. Zool., Lond.* 183: 311–328.

64. GRUBB, T. C. JR. 1974. Olfactory navigation to the nesting burrow in Leach's Petrel (*Oceanodroma leucorhoa*). *Anim. Behav.* 22: 192–202.
65. —— 1978. Weather-dependent foraging rates of wintering woodland birds. *Auk* 95: 370–376.
66. GYLLIN, R. KÄLLANDER, H. and SYLVÉN, M. 1977. The microclimate explanation of town centre roosts of Jackdaws *Corvus monedula*. *Ibis* 119: 358–361.
67. HARPER, W. G. 1960. An unusual indicator of convection. *Mar. Obs.* 30: 36–40.
68. HAYNES, V. M. 1980. Communal roosting by Wrens. *Brit. Birds* 73: 104–105.
69. HENTY, C. J. 1977. Thermal soaring of raptors. *Brit. Birds* 70: 471–475.
70. HENWOOD, K. and FABRICK, A. 1979. A quantitative analysis of the dawn chorus: temporal selection for communicatory optimization. *Amer. Nat.* 114: 260–274.
71. HEPPLESTON, P. 1971. The feeding ecology of Oystercatchers (*Haematopus ostralegus*) in winter in Northern Scotland. *J. Anim. Ecol.* 40: 651–672.
72. HICKLING, R. A. O. 1957. The social behaviour of gulls wintering inland. *Bird Study* 4: 181–192.
73. HOLDGATE, M. W. (ed.) 1971. *The Sea Bird Wreck in the Irish Sea Autumn 1969*. Nat. Environ. Res. Council, London.
74. HOUSTON, D. C. 1980. A possible function of sunning behaviour by griffon vultures *Gyps* spp., and other large soaring birds. *Ibis* 122: 366–369.
75. JENNINGS, A. R. 1961. An analysis of 1000 deaths in wild birds. *Bird Study* 8: 25–31.
76. JOHNSON, C. G. 1957. The vertical distribution of aphids in the air and the temperature lapse rate. *Q. J. Roy. Met. Soc.* 83: 194–201.
77. KEETON, W. T. 1974. The mystery of pigeon homing. *Sci. Amer.* 231 (6): 96–107.
78. KENDEIGH, S. C. 1952. Parental care and its evolution in birds. *Ill. Biol. Monogr.* 22: 1–356.
79. KENNEDY, R. J. 1969. Sunbathing behaviour of birds. *Brit. Birds* 62: 249–258.
80. —— 1970. Direct effects of rain on birds: a review. *Brit. Birds* 63: 401–414.
81. LACK, D. 1949. Family size in certain thrushes (Turdidae). *Evolution* 3: 57–66.
82. —— 1955. The summer movements of Swifts in England. *Bird Study* 2: 32–40.
83. —— 1958a. Weather movements of Swifts 1955–1957. *Bird Study* 5: 128–142.
84. —— 1958b. The return and departure of Swifts *Apus apus* at Oxford. *Ibis* 100: 477–502.
85. —— 1963. Migration across the southern North Sea studied by radar. Part 5. Movements in August, winter and spring, and conclusion. *Ibis* 105: 461–492.
86. —— 1965. *The Life of the Robin*. 4th ed. London.
87. —— 1966. *Population Studies of Birds*. Oxford.
88. —— 1973. *Swifts in a Tower*. (repr.) London.
89. —— and EASTWOOD, E. 1962. Radar films of migration over eastern England. *Brit. Birds* 55: 388–414.
90. LEACH, I. H. 1981. Wintering Blackcaps in Britain and Ireland. *Bird Study* 28: 5–14.
91. LEE, S. L. B. 1963. Migration in the Outer Hebrides studied by radar. *Ibis* 105: 493–515.
92. LOCKLEY, R. M. 1942. *Shearwaters*. London.
93. —— 1953. *Puffins*, London.
94. LÖHRL, H. 1976. Studies of less familiar birds. 179. Collared Flycatcher. *Brit. Birds* 69: 20–26.

95. MACDONALD, S. D. and PARMELEE, D. F. 1962. Feeding behaviour of the Turnstone in arctic Canada. *Brit. Birds* 55: 241–243.
96. MCILHENNY, E. A. 1940. Effects of excessive cold on birds in southern Louisiana. *Auk* 57: 408–410.
97. MCINTOSH, D. H. (Ed.) 1972. *Meteorological Glossary*. London.
98. MACLEAN, S. F. 1974. Lemming bones as a source of calcium for Arctic sandpipers (*Calidris* spp.). *Ibis* 116: 552–557.
99. MANIKOWSKI, S. 1971. The influence of meteorological factors on the behaviour of sea-birds. *Acta Zool. Cracov.* 16: 581–668.
100. MARISSENS, J. 1973. Exceptional European weather events in 1972. *Weather* 28: 347–350.
101. MARSHALL, A. J. 1959. Internal and environmental control of breeding. *Ibis* 101: 456–478.
102. MASON, B. J. 1979. The distinction between weather and climate. *Met. Mag.* 108: 211-212.
103. MCGOWAN, J. D. 1969. Starvation of Alaskan Ruffed Grouse and Sharp-tailed Grouse caused by icing. *Auk* 86: 142–143.
104. MEINERTZHAGEN, R. 1955. The speed and altitude of bird flight (with notes on other animals). *Ibis* 97: 81–117.
105. MES, R., SCHUCKARD, R. and WATTEL, J. 1978. Visdieven *Sterna hirundo* zoeken koelte. *Limosa* 51: 64–68.
106. MILLS, E. L. 1969. Hurricane 'Gladys' and its ornithological effect on the Maritime Provinces. *Nova Scot. Bird Soc. Newsletter* 11: 6–16.
107. MOREAU, R. E. 1972. *The Palaearctic-African Bird Migration Systems*. London.
108. MOTE, W. R. 1969. Turkey Vultures land on vessel in fog. *Auk* 86: 766–767.
109. MURRAY, R. 1966. A note on the large-scale features of the 1962/63 winter. *Met. Mag.* 95: 339–348.
110. —— 1977. The 1975/76 drought over the U.K. – hydrometeorological aspects. *Met. Mag.* 106: 129–145.
111. MURTON, R. K. 1965. *The Woodpigeon*. London.
112. MYTON, B. A. and FICKEN, R. W. 1967. Seed size preference in Chickadees and Titmice in relation to ambient temperature. *Wils. Bull.* 79: 319–321.
113. NELSON, J. B. 1965. The behaviour of the Gannet. *Brit. Birds* 58: 233–288, 313–336.
114. —— 1978. *The Gannet*. Berkhamsted.
115. NETHERSOLE-THOMPSON, D. 1966. *The Snow Bunting*. Edinburgh.
116. —— 1973. *The Dotterel*. London.
117. —— 1975a. Summer food and feeding habitat of the Greenshank. *Brit. Birds* 68: 243–245.
118. —— 1975b. *Pine Crossbills*. Berkhamsted.
119. —— and NETHERSOLE-THOMPSON, C. 1979. *Greenshanks*. Berkhamsted.
120. NEWTON, I. 1966. Fluctuations in the weights of Bullfinches. *Brit. Birds* 59: 89–100.
121. —— 1972. *Finches*. London.
122. —— 1979. *Population Ecology of Raptors*. Berkhamsted.
123. NISBET, I. C. T. 1962. South-eastern rarities at Fair Isle. *Brit. Birds* 55: 74–86.
124. —— 1963. American passerines in western Europe, 1951–62. *Brit. Birds* 56: 204–217.
125. —— and DRURY, W. H. 1968. Short-term effects of weather on bird migration: a field study using multivariate statistics. *Anim. Behav.* 16: 496–530.
126. NYE, P. A. 1964. Heat loss in wet ducklings and chicks. *Ibis* 106: 189–197.

127. O'CONNOR, R. J. 1977. Differential growth and body composition in altricial passerines. *Ibis* 119: 147–166.
128. OGILVIE, M. A. 1978. *Wild Geese*. Berkhamsted.
129. —— and ST. JOSEPH, A. K. M. 1976. Dark-bellied Brent Geese in Britain and Europe, 1955–76. *Brit. Birds* 69: 422–439.
130. OJANEN, M. 1979. Effect of a cold spell on birds in northern Finland in May 1968. *Ornis Fenn.* 56: 148–155.
131. OWEN, D. F. 1960. The nesting success of the Heron *Ardea cinerea* in relation to the availability of food. *Proc. Zool. Soc. Lond.* 133: 597–617.
132. PARRINDER, E. R. 1964. Little Ringed Plovers in Britain during 1960–62. *Brit. Birds* 57: 191–198.
133. —— and PARRINDER, E. D. 1975. Little Ringed Plovers in Britain in 1968–73. *Brit. Birds* 68: 359–368.
134. PENNYCUICK, C. J. 1969. The mechanics of bird migration. *Ibis* 11: 525–556.
135. —— 1972. Soaring behaviour and performance of some East African birds, observed from a motor glider. *Ibis* 114: 178–218.
136. PERRINS, C. M. 1970. The timing of birds' breeding seasons. *Ibis* 112: 242–255.
137. —— 1979. *British Tits*. London.
138. PHILLIPS, J. H. and LEE, the late S. L. B. 1966. Movements of Manx Shearwaters off Erris Head, Western Ireland, in the autumn. *Bird Study* 13: 284–296.
139. PIENKOWSKI, M. W., LLOYD, C. S. and MINTON, C. D. T. 1979. Seasonal and migrational weight changes in Dunlins. *Bird Study* 26: 134–148.
140. PLUMB, W. J. 1965. Observations on the breeding biology of the Razorbill. *Brit. Birds* 58: 449–456.
141. PORTER, R. and WILLIS, I. 1968. The autumn migration of soaring birds at the Bosphorus. *Ibis* 110: 520–536.
142. POTTS, G. R. 1969. The influence of eruptive movements, age, population size and other factors on the survival of the Shag '*Phalacrocorax aristotelis*'. *J. Anim. Ecol.* 38: 53–102.
143. PROWSE, A. D. 1966. Reactions of feeding Skylarks and Snipe to adverse weather. *Brit. Birds* 59: 434–435.
144. REYNOLDS, C. M. 1979. The heronries census: 1972–77 population changes and a review. *Bird Study* 26: 7–12.
145. RICHARDSON, W. J. 1972. Autumn migration and weather in eastern Canada – a radar study. *Amer. Birds* 26: 10–17.
146. —— 1976. Autumn migration over Puerto Rico and the western Atlantic – a radar study. *Ibis* 118: 309–332.
147. —— 1978. Timing and amount of bird migration in relation to weather – a review. *Oikos* 30: 224–272.
148. RIDER, G. C. and SIMPSON, J. E. 1968. Two crossing fronts on radar. *Met. Mag.* 97: 24–30.
149. ROTH, R. R. 1976. Effects of a severe thunderstorm on airborne ducks. *Wils. Bull.* 88: 654–656.
150. SAGE, B. L. 1969. Breeding biology of the Coot. *Brit. Birds* 62: 134–143.
151. SALOMONSEN, F. 1967. Migratory movements of the Arctic Tern (*Sterna paradisaea* Pontopp.) in the Southern Ocean. *Biol. Meddr. Dansk Vid. Selsk.* 24: 1–42.
152. SANDEMAN, G. L. 1974. A large movement of Kittiwakes in the Forth. *Scott. Birds* 8: 77–78.
153. SHELLARD, H. C. 1968. The winter of 1962–63 in the United Kingdom – a climatological survey. *Met. Mag.* 97: 129–141.

154. SHOOTER, P. 1970. The Dipper population of Derbyshire, 1958–68. *Brit. Birds* 63: 158–163.
155. —— 1978. Kingfisher diving through ice to catch fish. *Brit. Birds* 71: 130.
156. SIMMS, E. 1978. *British Thrushes*. London.
157. SIMPSON, J. E. 1967a. Swifts in sea-breeze fronts. *Brit. Birds* 60: 225–239.
158. —— 1967b. Aerial and radar observations of some sea-breeze fronts. *Weather* 22: 306–316.
159. SMITH, A. G. and WEBSTER, H. R. 1955. Effects of hail storms on waterfowl populations in Alberta, Canada – 1953. *J. Wildl. Mgmt.* 19: 368–374.
160. SMITH, L. P. 1979. The effect of weather on the incidence of botulism in waterfowl. *Agric. Meteorol.* 20: 483–488.
161. SNOW, D. W. 1955. The abnormal breeding of birds in the winter 1953/54. *Brit. Birds* 48: 120–126.
162. —— 1958. The breeding of the Blackbird *Turdus merula* at Oxford. *Ibis* 100: 1–30.
163. SPITZER, G. 1972. Jahreszeitliche Aspekte der Biologie der Bartmeise (*Panurus biarmicus*) *J. Orn.* 113: 241–275.
164. STANLEY, P. I. and MINTON, C. D. T. 1972. The unprecedented westward migration of Curlew Sandpipers in autumn 1969. *Brit. Birds* 65: 365–380.
165. STEEN, J. 1958. Climatic adaptation in some small northern birds. *Ecology* 39: 625–629.
166. SUZUKI, S., TANIOKA, K., UCHIMURA, S. and MARUMOTO, T. 1952. (The hovering height of Skylarks.) *J. Agr. Met., Japan.* 7: 149–151.
167. SVARDSON, G. 1957. The 'invasion' type of bird migration. *Brit. Birds* 50: 314–343.
168. SWANN, R. L. 1980. Fieldfare and Blackbird weights during the winter of 1978–79 at Drumnadrochit, Inverness-shire. *Ring. & Migr.* 3: 37–40.
169. TAYLOR, L. R. 1963. Analysis of the effect of temperature on insects in flight. *J. Anim. Ecol.* 32: 99–117.
170. THAKE, M A. 1977. Synoptic-scale weather and Honey Buzzard migration across the central Mediterranean. *Il-Merrill* (18): 19–25.
171. —— 1980. Gregarious behaviour among migrating Honey Buzzards *Pernis apivorus*. *Ibis* 122: 500–505.
172. THOMPSON, D. B. A. 1980. Unusual food of Shelduck. *Scott. Birds* 11: 83–84.
173. THRELFALL, W., EVELEIGH, E. and MAUNDER, J. E. 1974. Seabird mortality in a storm. *Auk* 91: 846–849.
174. THROWER, W. 1980. A wild goose storm disaster. *Norfolk Bird Rep.* 25: 102–104.
175. TUNNELL, G. A. 1964. The winter of 1962–63 and its effect upon British coastal waters. *Mar. Obs.* 34: 21–32.
176. UDVARDY, M. D. F. 1954. Summer movements of Black Swifts in relation to weather conditions. *Condor* 56: 261–267.
177. VERNON, J. D. R. 1969. Spring migration of the Common Gull in Britain and Ireland. *Bird Study* 16: 101–107.
178. WALLACE, J. M. AND HOBBS, P. V. 1977. *Atmospheric Science – an introductory survey*. London and New York.
179. WATSON, A. 1965. Research on Scottish Ptarmigan. *Scott. Birds* 3: 331–349.
180. —— 1972. The behaviour of the Ptarmigan. *Brit. Birds* 65: 6–26; 93–117.
181. —— 1980. Starving Oystercatchers in Deeside after severe snowstorm. *Scott. Birds* 11: 55–56.
182. WATSON, D. 1977. *The Hen Harrier*. Berkhamsted.

183. WEIR, D. and PICOZZI, N. 1975. Aspects of social behaviour in the Buzzard. *Brit. Birds* 68: 125–141.
184. WHITMORE, R. C., MOSHER, J. A. and FROST, H. H. 1977. Spring migrant mortality during unseasonable weather. *Auk* 94: 778–781.
185. WILLIAMSON, K. 1959. The September drift-movements of 1956 and 1958. *Brit. Birds* 52: 334–377.
186. —— 1961. The concept of 'Cyclonic Approach'. *Bird Migr.* 1: 235–240.
187. —— 1963. Movements as an indicator of population changes. *Bird Migr.* 2: 207–223.
188. —— 1965. *Fair Isle and its birds.* Edinburgh.
189. —— 1968. Goose emigration from western Scotland. *Scott. Birds* 5: 71–89.
190. —— 1969a. Habitat preferences of the Wren on English farmland. *Bird Study* 16: 53–59.
191. —— 1969b. Weather systems and bird movements. *Q. J. Roy. Met. Soc.* 95: 414–423.
192. WINSTANLEY, D., SPENCER, R. and WILLIAMSON, K. 1974. Where have all the Whitethroats gone? *Bird Study* 21: 1–14.
193. WOOD, N. A. 1974. The breeding behaviour and biology of the Moorhen. *Brit. Birds* 67: 104–115; 137–158.
194. WOODCOCK, A. H. 1975. Thermals over the sea and gull flight behaviour. *Boundary-layer Met.* 9: 63–68.
195. WYNNE-EDWARDS, V. C. 1962. *Animal Dispersion in Relation to Social Behaviour.* Edinburgh and London.

Additional references mentioned in the second edition are shown below.

196. ABLE, K. P., BINGMAN, V. P., KERLINGER, P. and GERGITS, W. 1982. Field studies of avian nocturnal migratory orientation II. Experimental manipulation of orientation in White-throated Sparrows (*Zonotrichia albicollis*) released aloft. *Anim. Behav.* 30: 768–773.
197. BAILLIE, S. R., CLARK, N. A. and OGILVIE, M. A. 1986. Cold weather movements of waterfowl and waders: an analysis of ringing recoveries. *Commissioned Report to N.C.C.*, B.T.O., Tring.
198. BAKER, J. K. and CATLEY, G. P. 1987. Yellow-browed Warblers in Britain and Ireland, 1968–85. *Brit. Birds* 80: 93–109.
199. BAKER, R. R. 1984. *Bird Migration: the solution of a mystery?* London.
200. COCHRAN, W. W. and KJOS, C. G. 1985. Wind drift and migration of thrushes: a telemetry study. *Ill. Nat. Hist. Surv.* 33: 297–330.
201. ELKINS, N. 1986. Vagrants and Saharan dust. *Brit. Birds* 79: 304–305.
202. —— 1987. Origin of Arctic Gulls in Britain and Ireland. *Brit. Birds* 80: 635–637.
203. FOLLAND, C. K., PALMER, T. N. and PARKER, D. E. 1986. Sahel rainfall and worldwide sea temperatures 1901–85. *Nature* 320: 602–607.
204. GUSTAFSON, T., LINDKVIST, B., GOTBORN, L. and GYLLIN, R. 1977. Altitudes and flight times for Swifts *Apus apus* L. *Orn. Scand.* 8: 87–95.
205. HARRIS, M. P. 1984. *The Puffin.* Calton.
206. —— and WANLESS, S. 1985. Fish fed to young Guillemots, *Uria aalge* and used in display on the Isle of May, Scotland. *J. Zool., Lond* (A) 207: 441–458.
207. MIKKOLA, H. 1983. *Owls of Europe.* Calton.

208. MILLSAP, B. A. and ZOOK, J. R. 1983. Effects of weather on accipiter migration in southern Nevada. *Rapt. Res.* 17: 43–56.

209. NISBET, I. C. T. 1983. Belly-soaking by incubating and brooding Common Terns. *J. Fld. Ornith.* 54: 190–192.

210. PENNYCUICK, C. H. 1982. The flight of petrels and albatrosses (Procellariiformes), observed in South Georgia and its vicinity. *Phil. Trans. R. Soc. Lond.* B 300: 75–106.

211. SIMMONS, K. E. L. 1986. *The sunning behaviour of birds.* Bristol.

212. THAKE, M. A. 1981. Falls of migrating Kestrels induced by a cold pool. *Riv. Ital. Orn.* 51: 241–247.

213. UNDERWOOD, L. A. and STOWE, T. J. 1984. Massive wreck of seabirds in eastern Britain, 1983. *Bird Study* 31: 79–88.

APPENDIX

Mean sea level pressures

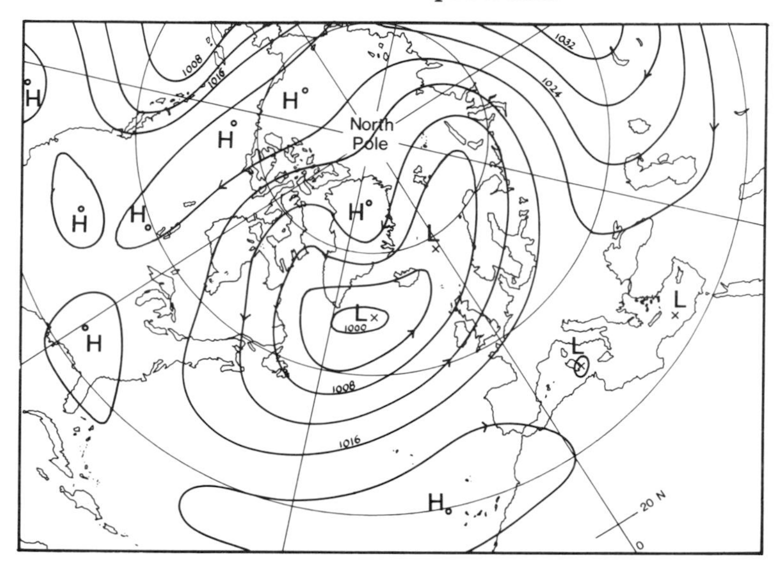

Fig. 47 Mean sea level pressure, January. Circulation at its strongest in winter hemisphere. Note intense Asian anticyclone and strong southwesterly flow over Atlantic and northwest Europe – creating relatively mild conditions in maritime regions.

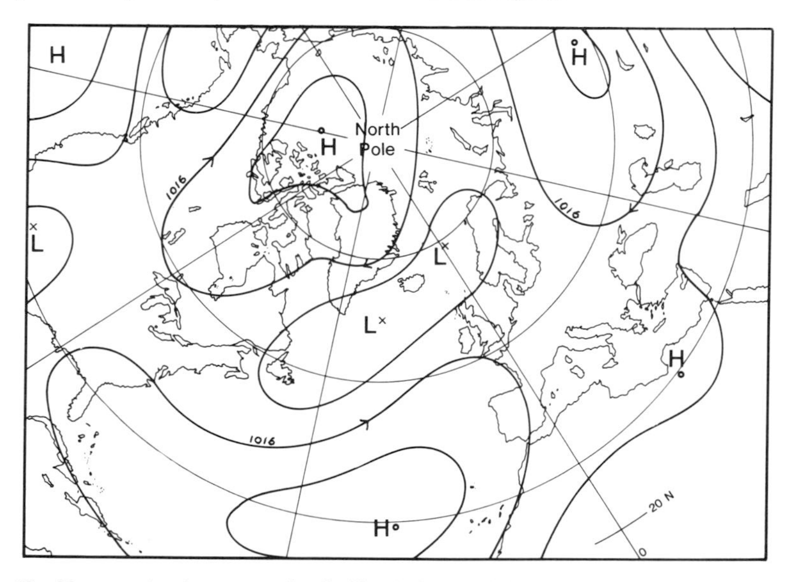

Fig. 48 Mean sea level pressure, April. Circulation considerably weaker than January. Ridges of high pressure over Europe facilitating northward migration. Similar situation in North America where axis of ridge over eastern seaboard 10° further south than in autumn – spring migrants use overland route to reduce risk of drift over sea in westerlies. Weak pressure distribution over polar migration routes facilitates passage of arctic species.

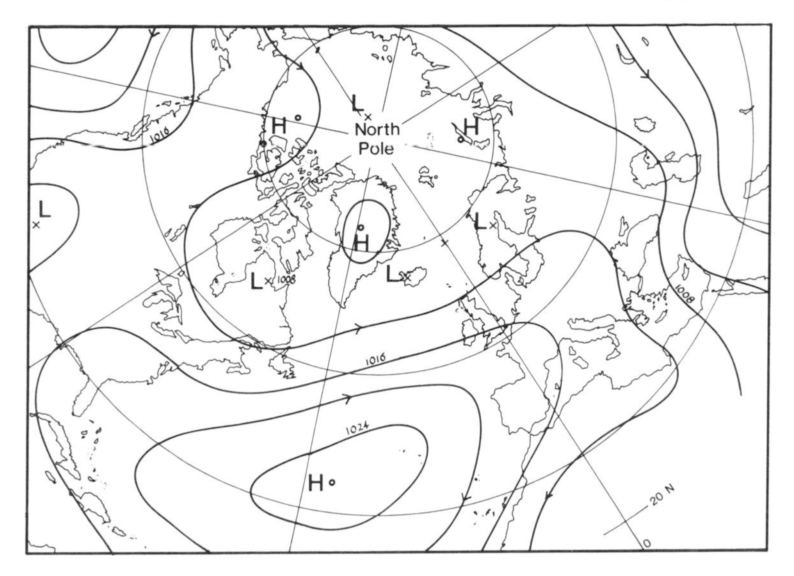

Fig. 49 Mean sea level pressure, July. Strong ridge over Europe.

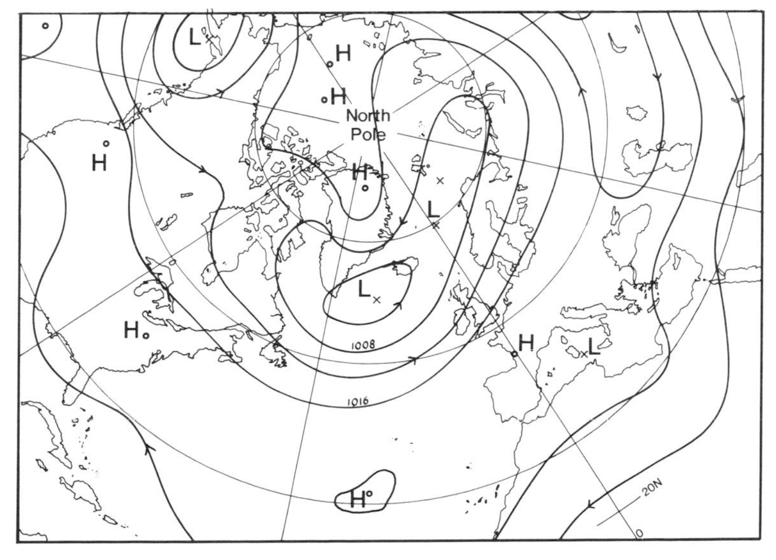

Fig. 50 Mean sea level pressure, October. Strengthening circulation as winter approaches (September circulation similar but weaker). Frequency of southwesterlies over Scandinavia implies mass migration on relatively few occasions. Anticyclone over Europe and in USA south of 45°N assists southbound migrants. Strong west to southwest flow over Atlantic north of 45°N – when this is further south than normal, more transatlantic vagrancy expected from west to east. Mean low southwest of Iceland not conducive to southeast passage from Greenland/Iceland, but secondary mean centre in Norwegian Sea implies a reasonable frequency of northwesterlies over route. WNW flow over eastern Canada indicative of polar airmasses assisting migration from this area. Intensifying anticyclone over central Asia with easterlies on southern flank assists westward dispersal of East Palearctic vagrants.

TABLE 1: *Some Symbols used on Synoptic charts.*

Symbol	Meaning	
L	Depression (Low) Centre	
H	Anticyclone (High) Centre	
	Cold front	Movement in direction of symbols
	Warm front	
	Occluded front (Occlusion)	
1008	Isobars, at 4 mbar intervals; the airflow is directed along the isobar with lower pressure to left of direction of flow (reversed in southern hemisphere)	
	Drizzle	
	Rain	
	Snow	
	Showers (qualified by precipitation symbol)	
	Fog	
	Thunderstorm	

Cloud cover in oktas (eighths of sky covered)

0	1	2	3	4	5	6	7	8	Sky or cloud invisible (usually obscured by fog)

(for wind symbols see Table 2)

TABLE 2: *Beaufort Wind Scale and Equivalent Wind Speeds (from McIntosh 1972).*[97]

Force	*Description*	*Symbol*	*Wind Speeds* knots (mean)	miles per hour (mean)	metres per second mean	metres per second limits
0	Calm		0	0	0	0·0–0·2
1	Light Air		2	2	0·8	0·3–1·5
2	Light Breeze		5	5	2·4	1·6–3·3
3	Gentle Breeze		9	10	4·3	3·4–5·4
4	Moderate Breeze		13	15	6·7	5·5–7·9
5	Fresh Breeze		19	21	9·3	8·0–10·7
6	Strong Breeze		24	28	12·3	10·8–13·8
7	Near Gale		30	35	15·5	13·9–17·1
8	Gale		37	42	18·9	17·2–20·7
9	Strong Gale		44	50	22·6	20·8–24·4
10	Storm		52	59	26·4	24·5–28·4
11	Violent Storm		60	68	30·5	28·5–32·6
12	Hurricane		(≥64)	(≥73)	≥32·7	

The plotted symbol on a synoptic chart points in the direction from which the wind is blowing. Every full barb = 10 knots; half barb = 5 knots; every filled triangle = 50 knots

e.g. = NW 20 knots (10 m/s)

TABLE 3: *Number of Wrens recorded at roost in Forest of Dean, Gloucestershire, 1979 (from Haynes 1980).*[68] *No counts were made during 18th–22nd February or 4th–5th March. Counts of arrivals as shown may have understated actual numbers.*

	Number counted	*Approx. temp. at roosting time (°C)*		*Number counted*	*Approx. temp. at roosting time (°C)*
30th Jan	'*c.* 2 dozen'	0	14th Feb	95	−2 to −3
31st Jan	41	0	15th Feb	96	−2 to −3
1st Feb	11	3	16th Feb	88	0
2nd Feb	63	0	17th Feb	80	1
3rd Feb	48	0	23rd Feb	40	4 (falling)
4th Feb	70	0	24th Feb	31	4 (falling)
5th Feb	70	0	25th Feb	35**	0 to 1
6th Feb	78	0	26th Feb	46	1 to 2 (falling)
7th Feb	72	1	27th Feb	27**	6
8th Feb	84	2	1st Mar	31	6·5
9th Feb	68	1	2nd Mar	8	11
10th Feb	87	0 to 1 (snowing)	3rd Mar	4	10 (windy and raining)
11th Feb	94	1 to 2	15th Mar	46	0 to 1 (snowing)
12th Feb	87	1	16th Mar	48	0 to 1 (snowing)
13th Feb	61*	3			

* = birds already leaving ** = bad visibility

TABLE 4: *Some avian responses to extreme temperatures.*

	Low temperature	*High temperature*
Maintenance of body temperature	Inactivity (conserves energy)	Panting, Gular fluttering (evaporative cooling)
	Shivering (increased heat production)	Exposure of skin (increased radiation loss)
	Intensive feeding	
	Covering of extremities	
	Control of blood heat to extremities	Increased blood heat flow to extremities
	Feather fluffing (increased insulation)	Sleeked plumage (reduced insulation)
	Huddling	Ruffling feathers in airflow
	Sheltering	Seeking shade, Body orientation (minimises radiation receipt)
Change of body temperature	Regulated hypothermia	
	Hypothermia	Hyperthermia
	Torpor	
	Hibernation	

TABLE 5: *Number of counts of passage of selected species of seabirds at Rattray Head, Aberdeenshire, during 1968–1971, according to weather conditions (from Elkins & Williams 1972).*[46]

	N–NW winds	*Onshore winds (NE–S)*	*Offshore winds (S–NW)*	*Anticyclonic light or variable winds*	*Counts tabulated*
Shearwater spp.	39	12	4	0	All seasons >10 per hour
Gannet	8	6	3	16	All seasons >100 per hour
Cormorant	0	0	1	9	Spring and autumn passage >50 per hour
Kittiwake	20	23	35	74	All seasons >250 per hour
Tern spp.	0	1	6	17	Autumn passage >50 per hour

Index of birds

Subject index

(For meteorological subjects, only main references are indexed)